Nuclear Medicine in Clinical Practice

100 self-assessment case studies

PAUL RYAN MSc, MRCP
Medway Hospital, Gillingham, UK

IGNAC FOGELMAN MD, FRCPS
Guy's Hospital, London, UK

B. LOUISE ROBINSON FRCR
Royal United Hospital, Bath, UK

Published by Chapman & Hall, 2–6 Boundary Row, London SE1 8HN, UK

Chapman & Hall, 2–6 Boundary Row, London SE1 8HN, UK

Chapman & Hall GmbH, Pappelallee 3, 69469 Weinheim, Germany

Chapman & Hall USA, 115 Fifth Avenue, New York, NY 10003, USA

Chapman & Hall Japan, ITP-Japan, Kyowa Building, 3F, 2-2-1 Hirakawacho, Chiyoda-ku, Tokyo 102, Japan

Chapman & Hall Australia, 102 Dodds Street, South Melbourne, Victoria 3205, Australia

Chapman & Hall India, R. Seshadri, 32 Second Main Road, CIT East, Madras 600 035, India

First edition 1997

Typeset by Type Study, Scarborough
Printed at the University Press, Cambridge

ISBN 0 412 45830 6

A Catalogue record for this book is available from the British Library

Preface

The role of nuclear medicine in the clinical management of patients is well established. One of its strengths is the ability to image the metabolic changes in response to a disease process and thus provide unique clinical information and high sensitivity for pathology. The high sensitivity is used, for example, with bone scanning for skeletal metastases where scan abnormalities appear before radiological change. The ability of nuclear medicine to image change in function and complement change in structure in disease states is seen with stress myocardial perfusion imaging in ischaemic heart disease. Here the reduction in blood flow due to arterial narrowing can be imaged non-invasively with thallium or other perfusion tracers. Another strength of nuclear medicine is the ability to image the whole body rapidly, and this is particularly useful for tumour localization.

There is, however, an ever increasing number of available nuclear medicine investigations and clinical applications. A trainee clinician requires knowledge and understanding of how various nuclear medicine investigations can be applied in medical practice. The purpose of this book is to provide examples of some of the more common imaging tests applied to clinical cases and covering the spectrum of nuclear medicine investigations currently undertaken.

Paul Ryan
Ignac Fogelman
B. Louise Robinson
Gillingham, London and Bath
June, 1997

Introduction

Examination candidates for FRCR, MRCP and equivalent postgraduate examinations overseas may only have had limited exposure to clinical nuclear medicine. A great deal can be learnt by reading standard texts, but these rarely give a feel for the use of nuclear medicine techniques in real-life clinical practice. This book has been produced to illustrate the role of nuclear medicine in day-to-day clinical decision-making. One hundred typical nuclear medicine cases are presented and the reader is invited to decided on the scan abnormality and its clinical significance. This format has been chosen to stimulate thought and maintain the reader's interest. The range of studies reflects current clinical practice and the type of cases that may be seen in examinations. The cases have deliberately not been arranged in a systematic manner so as to achieve some of the surprise of the examination format. With many of the cases we have included further reading lists for those who wish to read around the subject. There is an index at the back for those who wish to increase their knowledge in a more systematic manner.

Case 1

HISTORY

An 80-year-old woman presents with low back pain. A bone scan was performed and the posterior lumbar spine view is shown.

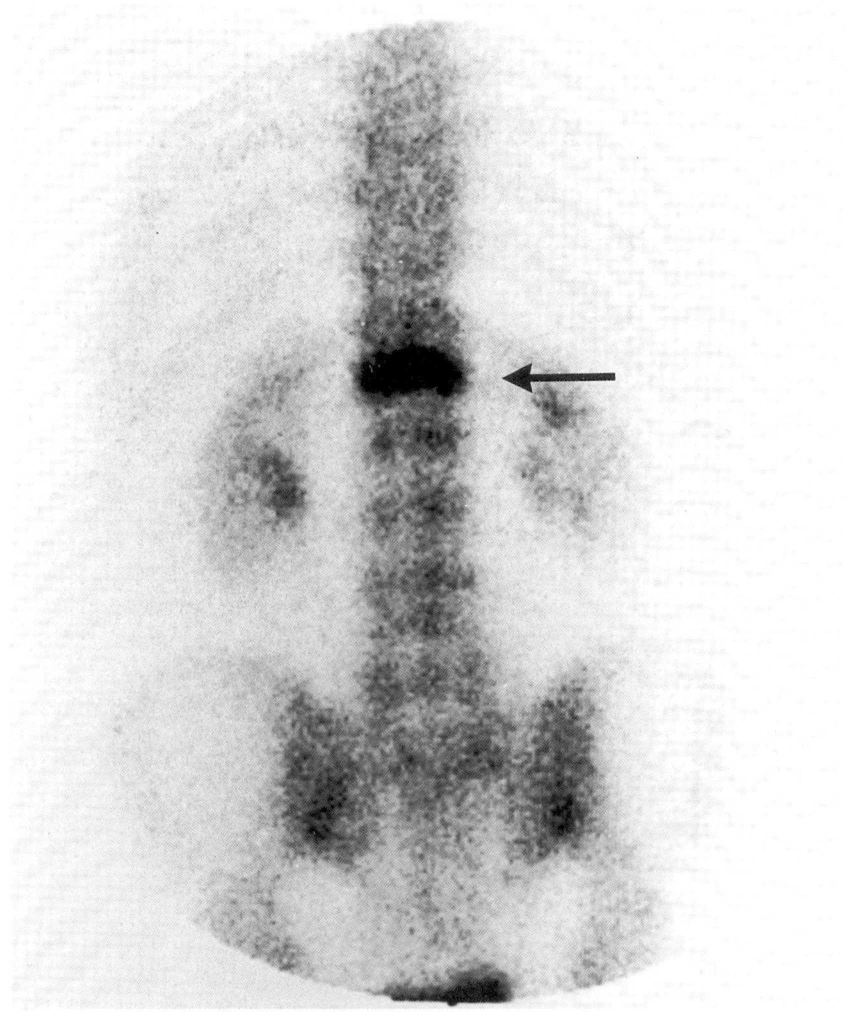

QUESTIONS

Q1. What does the arrowed abnormality suggest?
Q2. List three possible causes.

ANSWERS

Q1. Vertebral collapse.
Q2. Osteoporosis, trauma, malignancy.

TEACHING POINTS

1. Linear increased activity almost anywhere in the skeleton implies a fracture.
2. With a fracture, activity typically fades over six to nine months and intensity can be used as a guide to how recently a fracture occurred.
3. Multiple sites of linearly increased tracer uptake in the spine of varying intensity (with no significant lesions elsewhere throughout the skeleton) strongly supports a diagnosis of osteoporosis.

FURTHER READING

Ryan, P.J. and Fogelman, I. (1994) The role of nuclear medicine in orthopaedics. *Nuc. Med. Comm.*, **15**, 341–60.

Case 2

HISTORY

A 70-year-old woman presents with rib pain. A posterior chest bone scan image is shown.

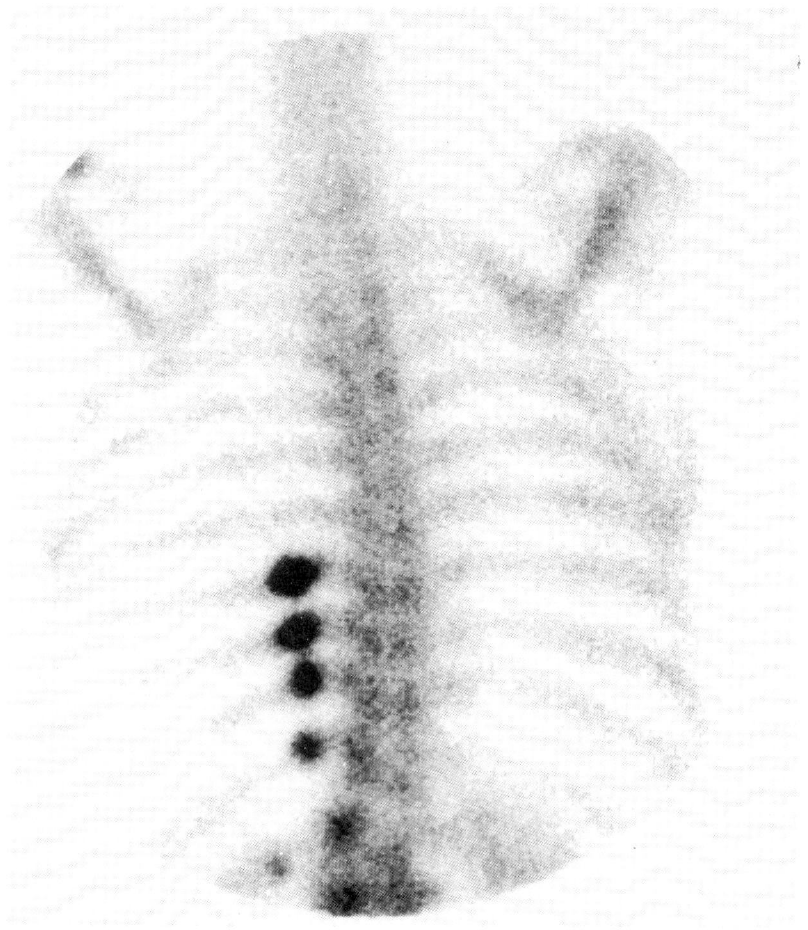

QUESTIONS

Q1. Describe the abnormalities in the ribs.
Q2. What is the most likely cause?

ANSWERS

Q1. Focal activity in posterior ribs in a linear pattern.
Q2. Trauma.

TEACHING POINT

Traumatic rib fractures commonly produce foci on a bone scan in a linear pattern. Randomly distributed rib activity in association with lesions at other sites or extending along ribs is likely to be metastatic.

Case 3

HISTORY

A 9-year-old girl presented with a urinary tract infection. A large left hydronephrosis was noted on a renal ultrasound, and the possibility of pelvi ureteric junction (PUJ) obstruction was considered. A MAG 3 was performed. Images are shown at 60 sec, 2 min, 5 min, 10 min, 15 min, 20 min and 25 min following injection, with the last images post micturition. The time–activity curve is shown (left kidney bold, right faint).

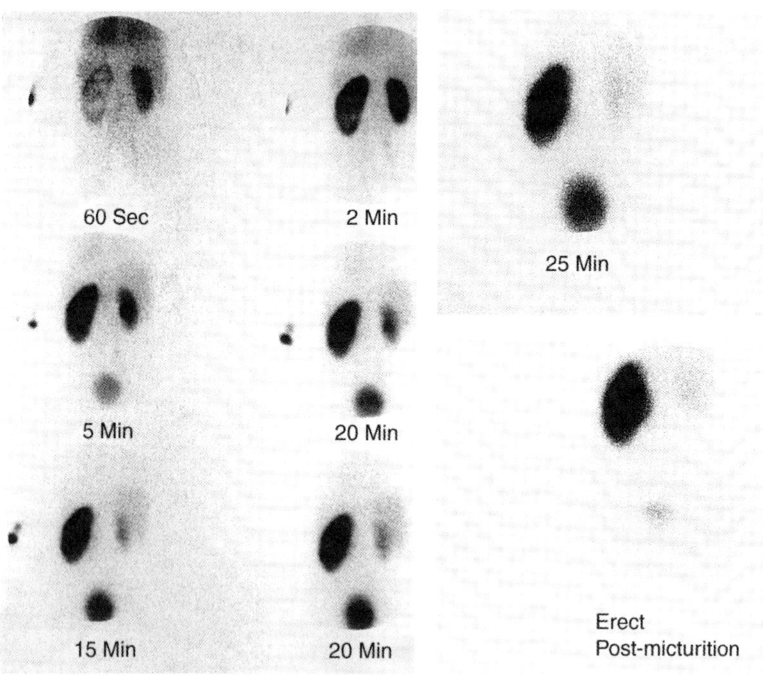

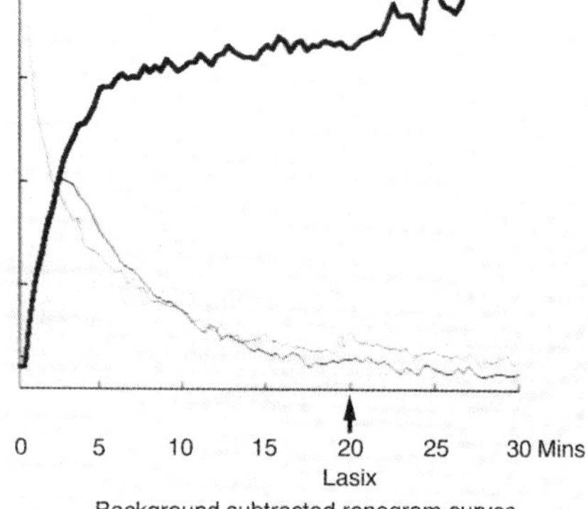

QUESTIONS

Q1. What are the abnormalities?

Q2. What does this imply?

ANSWERS

Q1. There is an enlarged left kidney with good function. There is slow transit and an accumulation of tracer into a dilated renal pelvis. There is no drainage following frusemide.

Q2. Left PUJ obstruction.

TEACHING POINT

MAG 3 or DTPA can be used as dynamic renal tracers to investigate PUJ obstruction. MAG 3 is often used in preference to DTPA as a dynamic renal tracer in children because of the greater extraction by the kidney; MAG 3 is filtered by the glomerulus and secreted by the tubules whereas DTPA is only filtered. MAG 3 produces a higher count rate resulting in better image contrast and count statistics.

Case 4

HISTORY

A 70-year-old man with carcinoma of the prostate. A whole body bone scan is displayed.

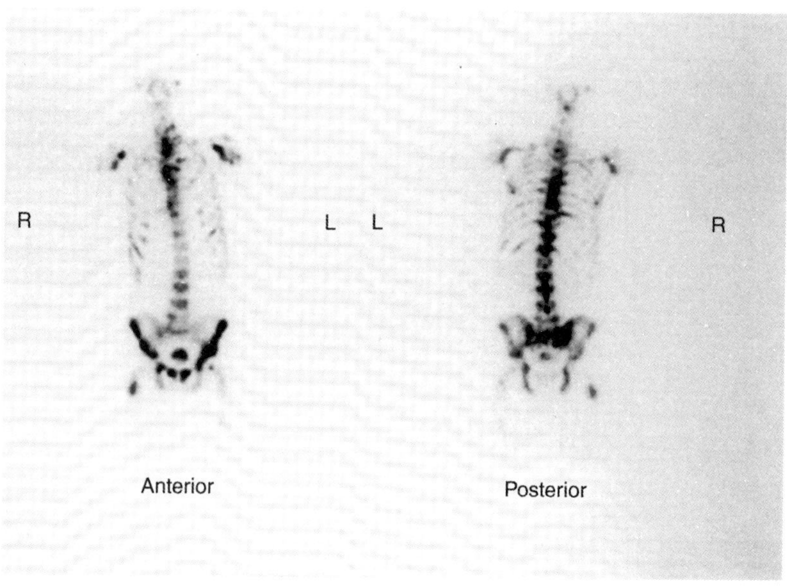

QUESTIONS

Q1. What are the findings?
Q2. What does this imply?

ANSWERS

Q1. Multiple foci of increased tracer uptake in the axial skeleton.
Q2. Widespread metastatic disease of the skeleton.

TEACHING POINTS

Note absent renal activity in this patient, implying unusually avid tracer uptake by the skeleton. This is approaching the situation of a super scan of malignancy where there is diffuse avid uptake by the axial skeleton and absent renal and sometimes bladder activity. This can be mistaken for a normal study by the inexperienced. A super scan can also be seen in metabolic bone disease, but this can generally be distinguished from malignancy.

Case 5

Anterior abdominal gallium scan images are shown in a patient with pyrexia of unknown origin (PUO), four days following injection.

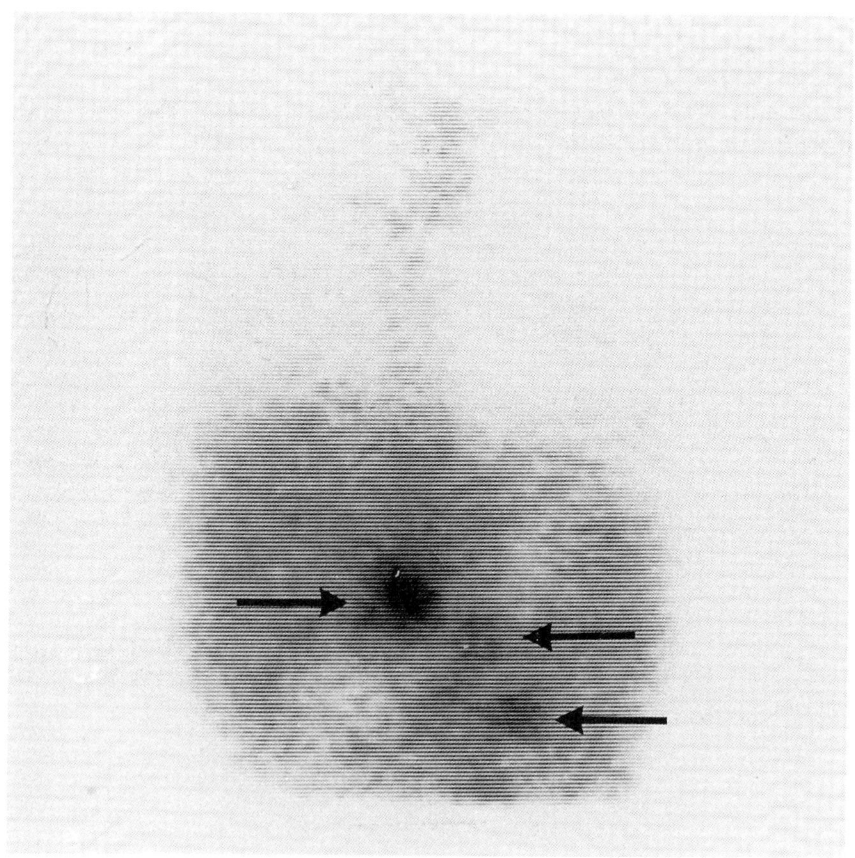

96 hours post-injection.

QUESTIONS

Q1. What abnormalities (arrowed) are shown?
Q2. List four possible causes.

ANSWERS

Q1. Focal increased tracer uptake in central anterior abdomen.
Q2. Focal uptake on gallium images could be explained by lymphoma, tuberculosis, pyogenic infection, or possibly sarcoid. This patient had a Hodgkin's lymphoma.

TEACHING POINTS

1. Normal sites of gallium uptake include liver, bowel and bone marrow. A little uptake is seen in lacrimal glands, salivary glands, spleen and sometimes breast. Gallium imaging is particularly useful for the investigation of PUO as it shows uptake in a wide range of conditions. Gallium imaging in lymphoma is most useful for the detection of a recurrence where computed tomography (CT) or magnetic resonance imaging (MRI) findings following radiotherapy may be difficult to distinguish from tumour. The sensitivity and specificity of gallium for the detection of lymphoma has been increased to over 90% with single photon emission computed tomography (SPECT) imaging.
2. Because of physiological excretion of tracer via the biliary system into bowel, detection of intra-abdominal pathology may be problematic. Imaging over several days can help distinguish 'normal' bowel activity from pathology. In some centres a laxative is given prior to the study.

FURTHER READING

Kortolzoglu, L., Heh, S.D.J., Portlock, C. *et al.* (1992) Validation of gallium 67 citrate single photon emission computed tomography in biopsy confirmed residual Hodgkins disease in the mediastinum. *J. Nuc. Med.*, **33**, 345–50.

Case 6

HISTORY

A 72-year-old man presents with back pain, pyrexia and raised erythrocyte sedimentation rate. A bone scan was performed. The posterior spine image is shown.

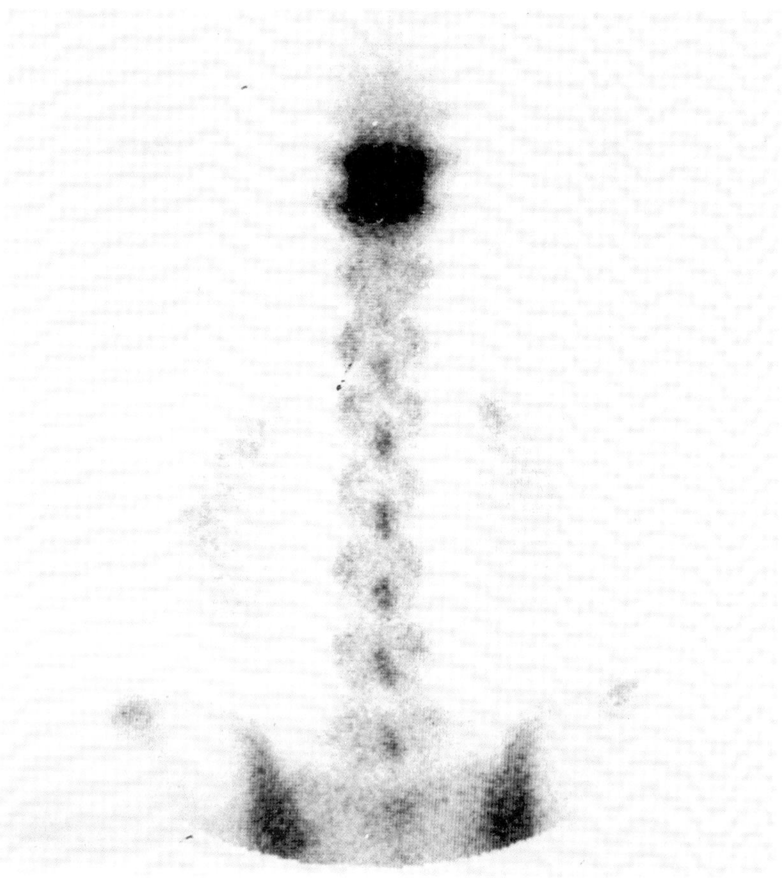

QUESTIONS

Q1. What abnormality is visualized?
Q2. What is the most probable cause?

ANSWERS

Q1. Increased activity T9/T10 involving two adjacent vertebrae and the intervening disc.
Q2. Discitis with secondary osteomyelitis.

TEACHING POINTS

1. Intense increased activity involving two vertebrae and adjacent joint space is typical for discitis. A gallium scan was used to confirm the diagnosis and is shown below. Gallium is preferable to labelled white cell scans in the spine because reduced uptake as well as increased uptake can be found in osteomyelitis using the latter agent. There are, however, many causes of reduced uptake on a labelled white cell scan, e.g. marrow replacement due to a metastasis.
2. Note the different pattern of bone scan uptake in this patient compared with vertebral fracture where there is intense uptake in a single vertebral body.

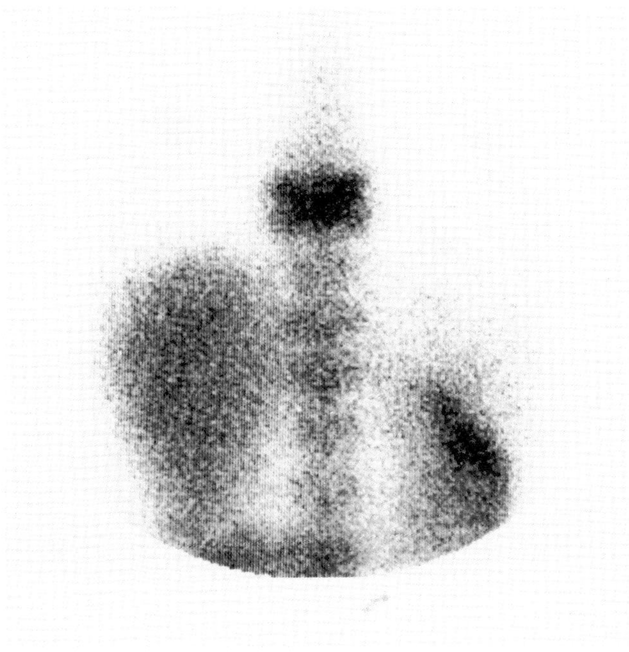

Gallium image of the same patient showing uptake at the site of suspected osteomyelitis.

Case 7

HISTORY

A 40-year-old man presents with long-standing abdominal pain and diarrhoea. One- and four-hour abdominal views were obtained following injection of ^{99m}Tc HMPAO (hexamethylpropyleneamine oxime) white cells.

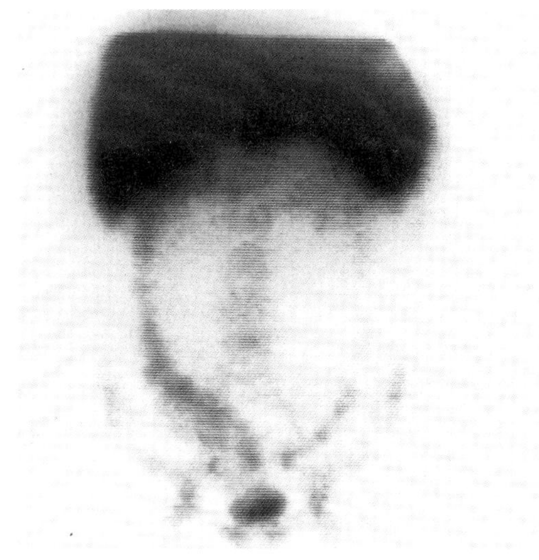

One-hour view.

Four-hour view.

QUESTIONS

Q1. What abnormality is shown?
Q2. What is the most likely cause?

ANSWERS

Q1. Increased uptake involving most of the large bowel, with an altering pattern of tracer activity between one and four hours.
Q2. Ulcerative colitis.

TEACHING POINTS

Labelled white cell (WCC) scans are an extremely sensitive and specific method for the diagnosis of inflammatory bowel disease. Uptake at one hour on Tc HMPAO-labelled WCC scan is almost always abnormal, although non-specific bowel activity may be seen at four hours. With inflammatory bowel disease, activity moves along the bowel, and this can be detected on later images.

Features that are more likely to suggest Crohn's disease are involvement of small bowel and skip lesions.

Case 8

HISTORY

On CT scan, a 52-year-old man was noted to have a small mass in the left posterior abdomen. He had a previous history of abdominal trauma. A diagnostic nuclear medicine scan was performed.

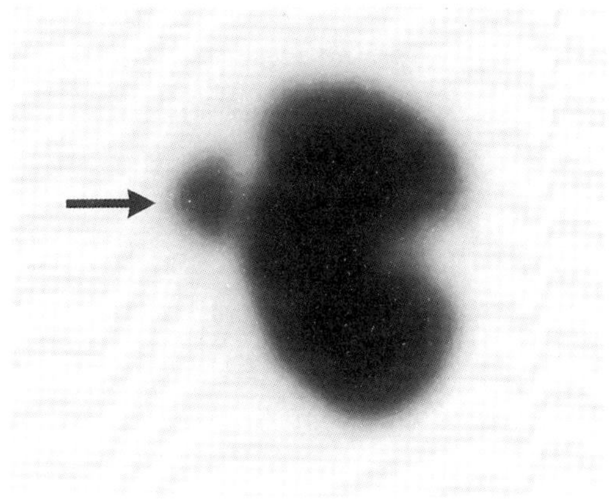

QUESTIONS

Q1. What scan was performed?
Q2. What does the arrowed abnormality show?

ANSWERS

Q1. A denatured red cell scan.
Q2. Splenunculus.

TEACHING POINT

A denatured red cell scan is specific for splenic tissue and can be used to identify a splenunculus. The spleen in the present study is bilobed, which is also abnormal.

Case 9

HISTORY

A patient was investigated for right upper quadrant pain.

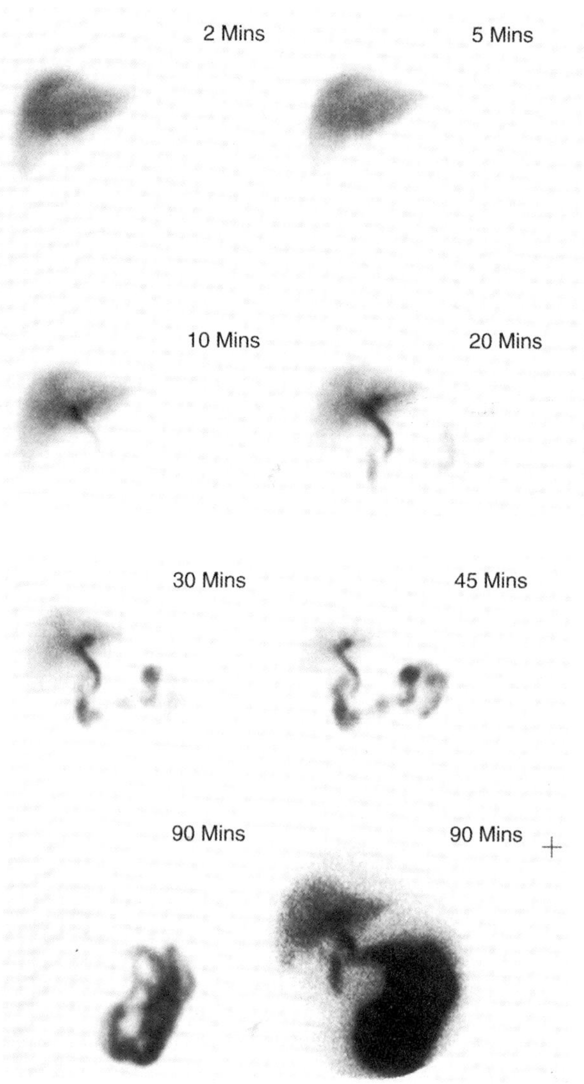

QUESTIONS

Q1. What is the scan?
Q2. What is the agent used?
Q3. What abnormality is shown?

ANSWERS

Q1. Liver and biliary tree imaging.
Q2. Tc labelled imino diacetic acid (HIDA).
Q3. Non-functioning gallbladder indicative of cholecystitis.

TEACHING POINTS

1. A Tc-HIDA study assesses cystic duct patency. Filling of the gallbladder within 30 minutes reliably excludes cholecystitis as the cause of the patient's pain. However, non-visualization of the gallbladder, although indicative of cholecystitis, is **not** diagnostic.
2. If a gallbladder is not visualized at 30 minutes following injection, the study should be continued until four hours because late filling may occur in chronic cholecystitis.

FURTHER READING

Water, A. and Kalff, V. (1991) Hepatobiliary imaging. *Curr. Op. Radiol.*, **3**, 851–8.

Case 10

HISTORY

A 15-year-old Nigerian girl presents with a painful right lower thigh. Blood pool and three-hour anterior delayed images of the legs above and below the knee are shown.

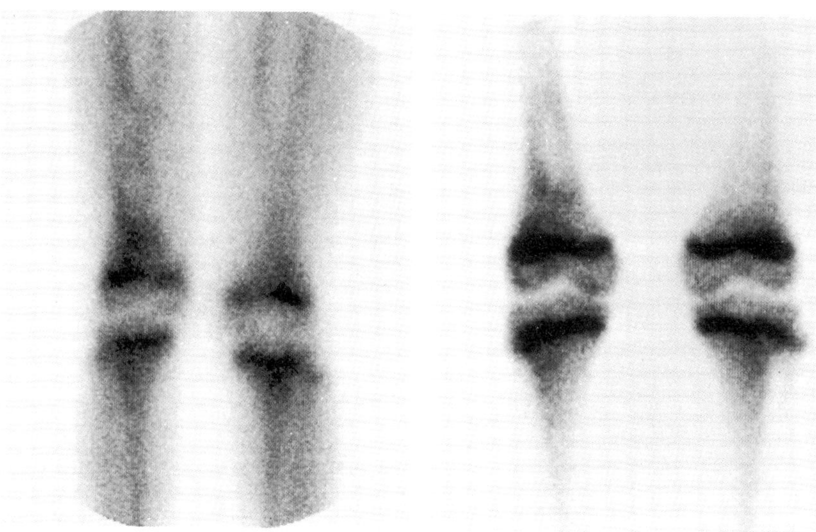

Blood pool. Three-hour anterior delayed view.

QUESTIONS

Q1. What abnormality is shown in the right lower femur?
Q2. What is the likely cause?

ANSWERS

Q1. Hyperaemia right lower femur on blood pool images, with absent activity on delayed views.
Q2. Bone infarct due to sickle cell disease.

TEACHING POINTS

1. In a sickle patient one may also see splenic uptake on the bone scan due to splenic infarction (arrowed). The mechanism of this uptake is unknown, but may be due to microcalcification. Activity may also be increased in the proximal long bones due to marrow hyperplasia (arrowed).
2. In the initial stages of infarction there is reduced bone scan activity, followed by increased uptake that then fades. The separation of bone infarction from osteomyelitis in sickle cell disease may therefore be difficult, and in patients with symptoms for more than a week a combined bone scan and gallium scan is recommended.

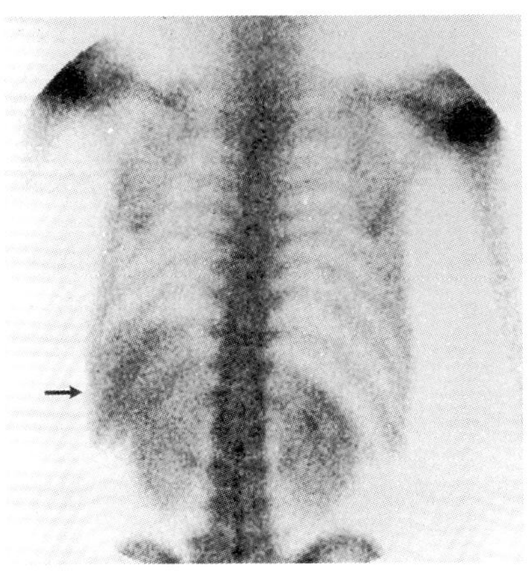

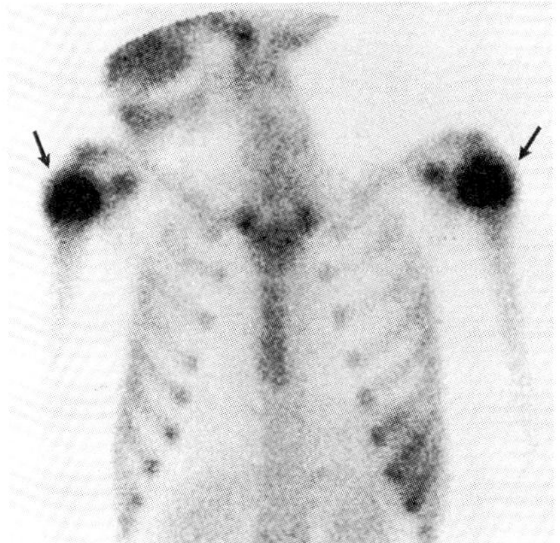

FURTHER READING

Amundsen, R.R., Siegal, M.J. and Siegal, B.A. (1984) Osteomyelitis and infarction in sickle cell haemoglobulinopathies: differentiation of combined technetium and gallium scintigraphy. *Radiology*, **153**, 807–12.

Case 11

HISTORY

A 34-year-old man presented with a urinary tract infection. A DMSA study was arranged. Anterior, posterior and left posterior oblique images are displayed.

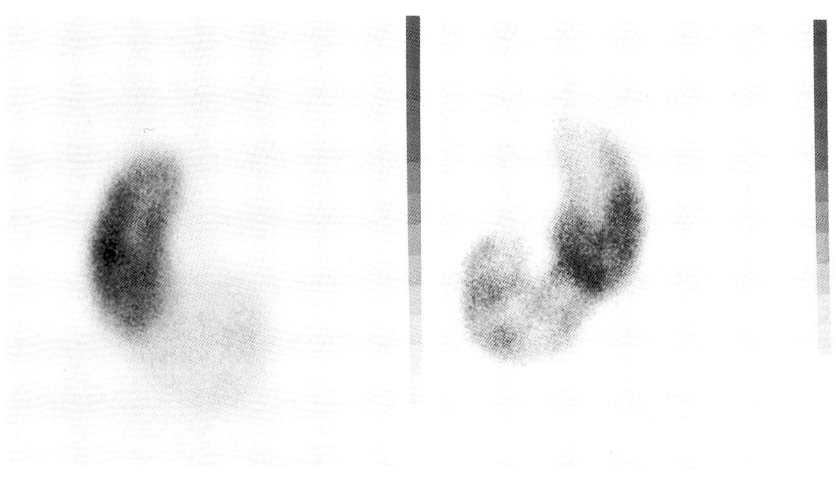

Posterior view. Anterior view.

Left posterior oblique.

QUESTIONS

Q1. What is the diagnosis?
Q2. How can a DMSA assist the clinician in this situation?

ANSWERS

Q1. Crossed fused renal ectopia.
Q2. The DMSA scan in this patient was useful firstly to make the diagnosis, and secondly to assess the function of the crossed fused kidney.

TEACHING POINTS

Renal images with ^{99m}Tc dimercaptosuccinic acid (DMSA) are obtained three hours after injection with most of the tracer fixed in the renal tubules. It is useful to assess the relative function of the two kidneys, to detect congenital anomalies, and to detect renal scarring. Other uses include the diagnosis of acute pyelonephritis, renal infarct, and renal artery stenosis (especially in children).

Case 12

HISTORY

A 55-year-old man developed acute abdominal pain five days following a right-sided renal transplant. Thirty-second and five-minute anterior views of the abdomen and pelvis are displayed following injection of ^{99m}Tc DTPA.

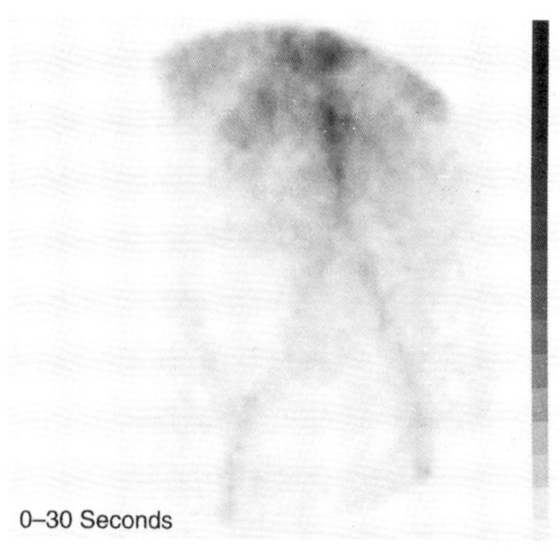

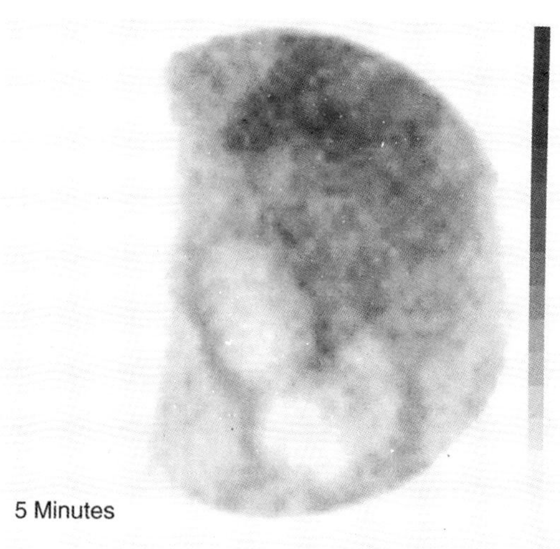

QUESTIONS

Q1. What is the abnormality?
Q2. What is the diagnosis?

ANSWERS

Q1. Absent perfusion and photon deficient area at site of renal transplant.
Q2. Renal artery thrombosis.

TEACHING POINT

An avascular renal transplant can be easily and rapidly detected with DTPA imaging. However, appearances can be due to:

- arterial thrombosis
- end-stage rejection
- hyperacute rejection
- venous occlusion.

These are indistinguishable on the scan.

Case 13

HISTORY

A six-year-old boy with left pulmonary artery stenosis underwent perfusion and ventilation lung scintigraphy after becoming increasingly short of breath.

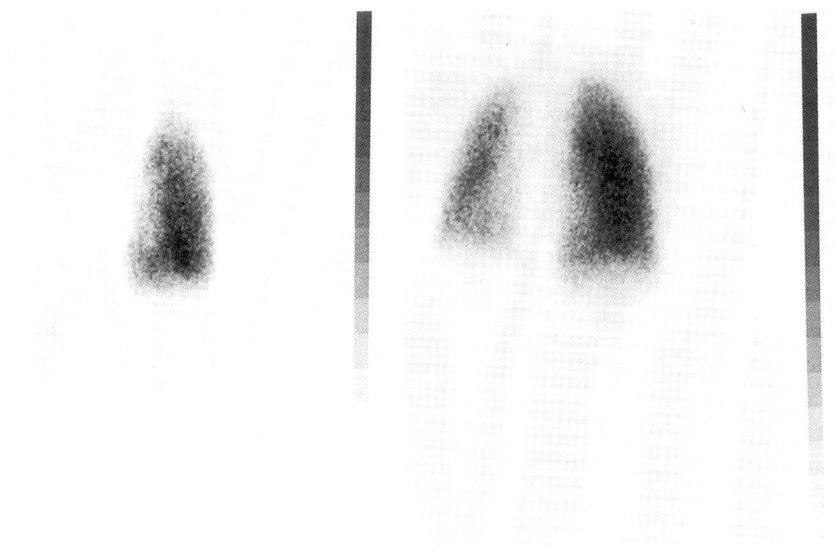

Perfusion scan. Ventilation scan.

QUESTIONS

Q1. What are the principal abnormalities?
Q2. How might you explain these?

ANSWERS

Q1. Absent perfusion and reduced ventilation in the left lung. There is also activity in the kidneys on perfusion imaging.

Q2. Changes in the left lung were due to pulmonary artery stenosis. Activity in the kidneys was due to patent foramen ovale, allowing the perfusion tracer to enter the systemic circulation (due to right to left shunt).

Case 14

HISTORY

A 34-year-old woman presents with pleuritic chest pain and dyspnoea: she is taking the oral contraceptive pill. Perfusion and ventilation lung scan images were obtained.

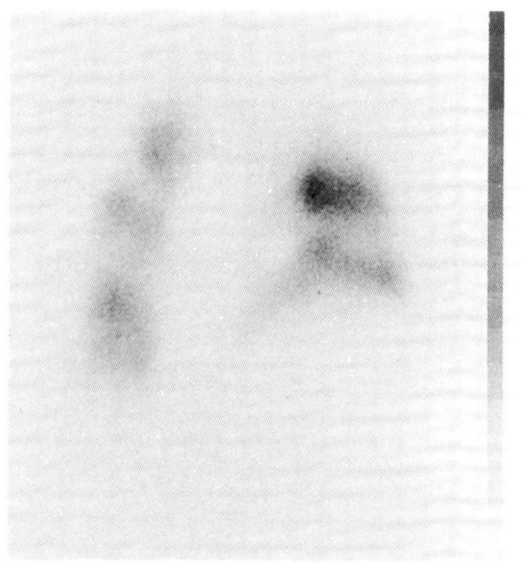

Right posterior oblique perfusion.

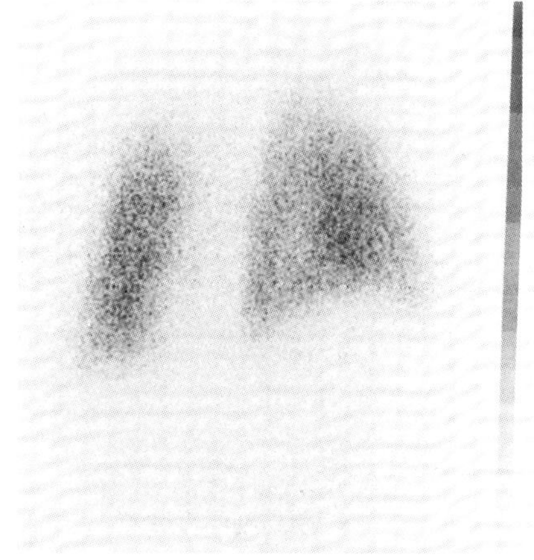

Right posterior oblique ventilation.

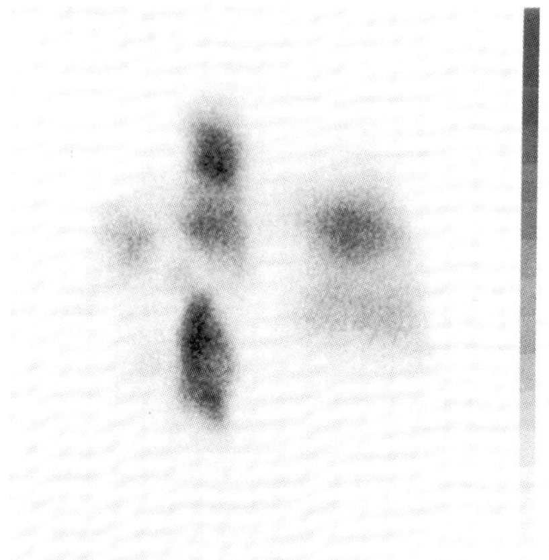

Left posterior oblique perfusion.

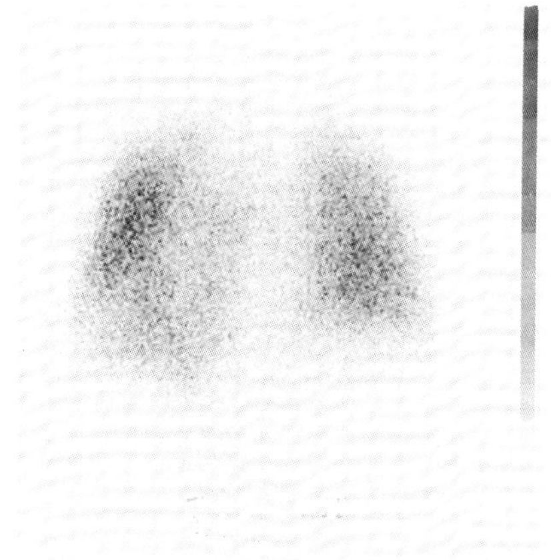

Left posterior oblique ventilation.

QUESTIONS

Q1. What abnormalities are demonstrated?
Q2. What agents could have been used for performing this study?

ANSWERS

Q1. Multiple segmental perfusion defects with normal ventilation. Classic appearances of pulmonary emboli.

Q2. Perfusion imaging is performed with ^{99m}Tc-labelled microspheres or macroaggregates of albumin (MAA). Ventilation is usually performed with krypton gas or ^{99m}Tc DTPA aerosol.

TEACHING POINTS

1. In patients with severe pulmonary hypertension, it may be unwise to perform perfusion imaging as the patient may be tipped into respiratory failure by temporary blockage of some pulmonary arterioles by the perfusion agent. This situation is rare, however, and it is customary to reduce the MAA dose by half if the study is required.
2. Lung scans may be performed during pregnancy and in lactating mothers, but lower doses are commonly used and breast-feeding should be suspended for 12 hours.

Case 15

HISTORY

A 44-year-old woman is admitted with acute renal impairment. A DTPA study is performed. Posterior images are displayed at 30 seconds, one minute, five minutes, 10 minutes and 30 minutes following injection.

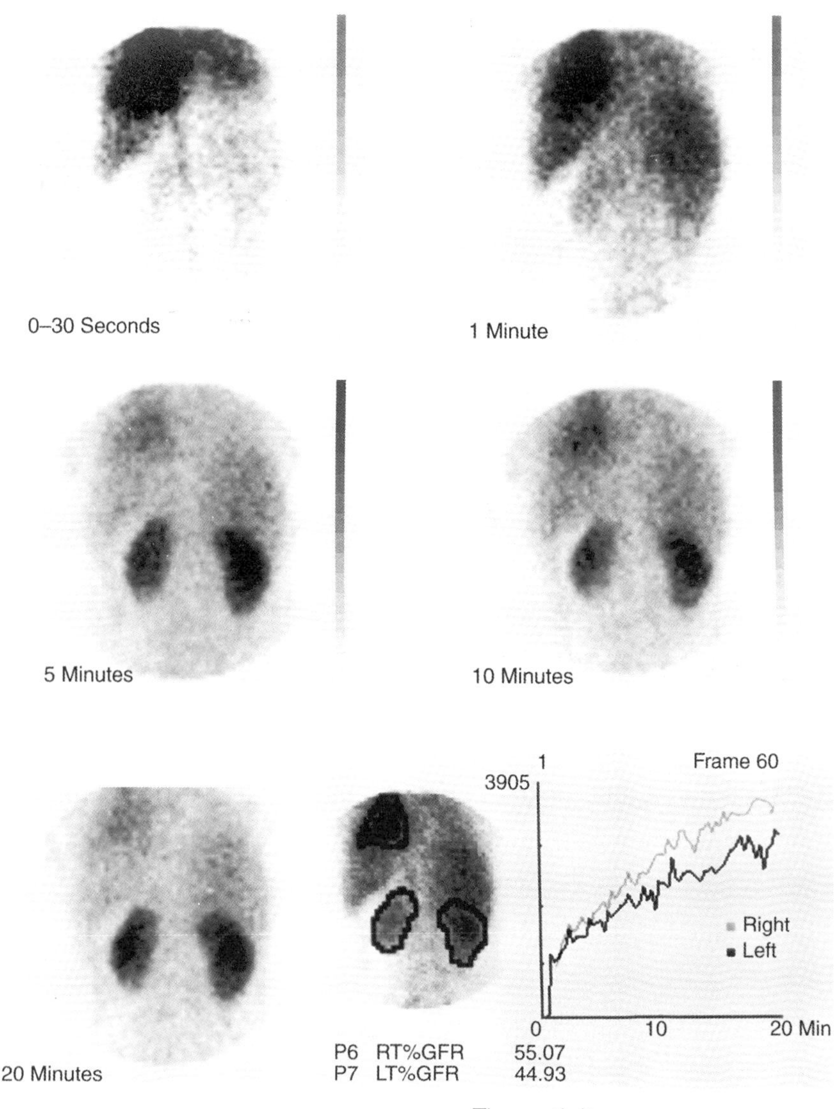

Time activity curve.

QUESTIONS

Q1. Describe the abnormalities in this study.
Q2. How would this assist the clinician in managing this patient?

ANSWERS

Q1. There is impaired perfusion of both kidneys with gradual accumulation of tracer in the parenchyma without significant excretion. These changes are typical of intrinsic renal failure.

Q2. Intrinsic renal failure can be separated from acute tubular necrosis (ATN) by a DTPA study. In ATN, there is typically normal perfusion and a blood pool, but subsequently there is loss of tracer from the kidney ('fading' blood pool image) without excretion as it diffuses from the vascular to the extravascular space.

TEACHING POINT

Renal outflow obstruction can produce the same renogram appearances in this study, although tracer uptake is usually seen in the collecting systems. A DTPA scan cannot be used as a substitute for renal ultrasound to exclude obstruction.

Case 16

HISTORY

A 60-year-old woman presents with right hip pain following a fall. Blood pool and delayed anterior pelvis images are shown.

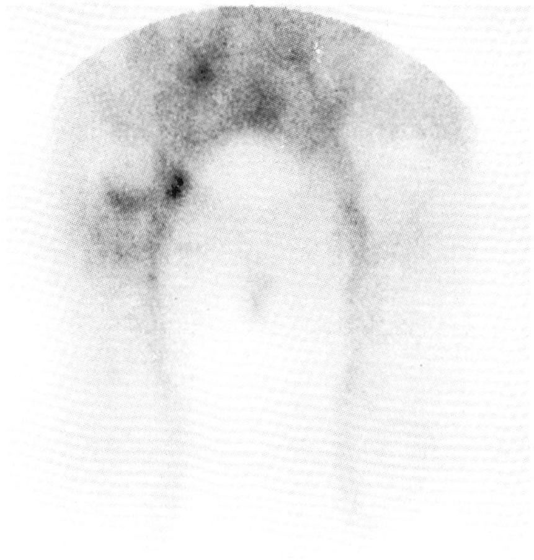

Blood pool.

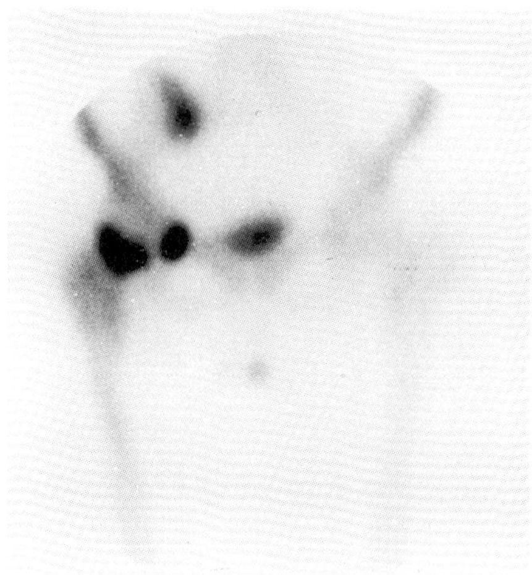

Delayed image, three hours post-injection.

QUESTIONS

Q1. What abnormalities are seen?
Q2. What is the probable cause?

ANSWERS

Q1. Linear increased activity right femoral neck, superior pubic rami and sacroiliac joint.
Q2. Fractures of femoral neck and in addition pelvic fracture.

TEACHING POINTS

1. The bone scan is frequently a useful test to exclude or confirm a fracture where one is suspected clinically but x-rays are normal. Where the bone scan is positive, x-rays will subsequently identify the fracture.
2. Recent fractures will generally show intense uptake on both blood pool and delayed images.

Case 17

HISTORY

A 50-year-old woman has a swollen left leg and a normal venogram. A radionuclide investigation was performed with anterior images of the feet and lower legs straight after injection and 30 minutes later.

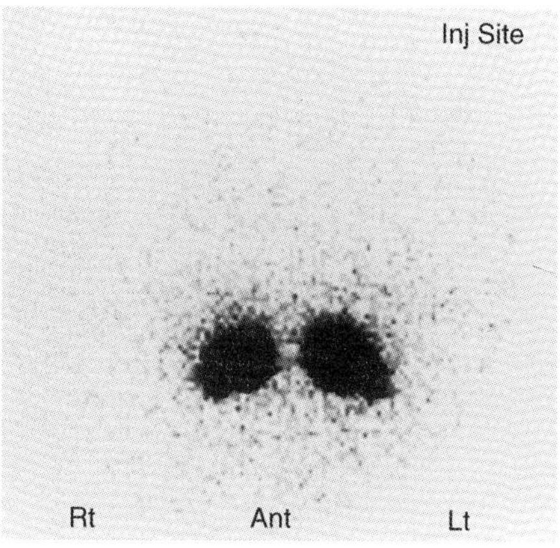

Immediately post-injection.

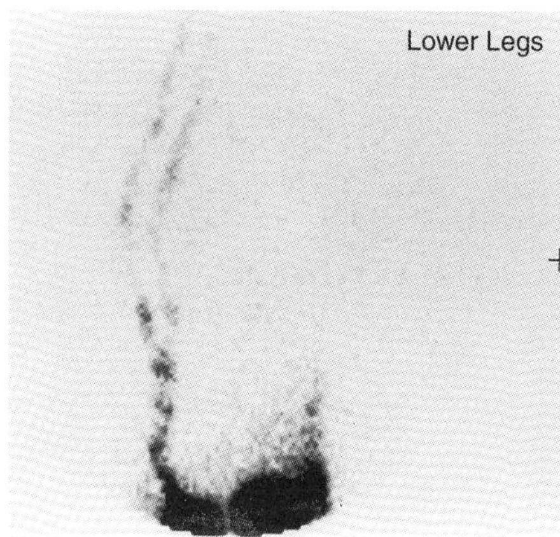

30 minutes post-injection.

QUESTIONS

Q1. What scan is this?
Q2. What agent is used?
Q3. What does it show?
Q4. What are the possible causes?

ANSWERS

Q1. Lymphoscintigram.
Q2. ^{99m}Tc nanocolloid.
Q3. Delayed passage of tracer via lymphatics of left leg.
Q4. Lymphatic obstruction may be:
- congenital, e.g. Klippel–Trenaunay syndrome
- primary non-hereditary
- primary hereditary (Milroy's disease)
- primary lymphoedema praecox, developing after one year (Meige's disease)
- acquired – filariasis
  pelvic infection/tumour
  trauma.

TEACHING POINT

Lymphoscintigraphy is easy to perform, involving only the injection of a small amount of ^{99m}Tc nanocolloid subcutaneously into the web space of each foot. There are no specific contraindications. The lymphatics are normally visualized within half an hour of injection, and groin lymph nodes within one hour. Assessment of the lymph nodes can be made, although variation in normal anatomy makes it difficult. Markedly asymmetrical nodes, focal defects, ballooning or mottled appearances of nodes are considered abnormal.

FURTHER READING

Intenso, C.M., Desai, A.G., Kim, S.G. *et al.* (1989) Lymphedema of the lower extremities: evaluation of microcolloidal imaging. *Clin. Nuc. Med.*, **14**, 101–7.

Case 18

This whole body radioiodine scan was performed in a 26-year-old woman with known thyroid carcinoma.

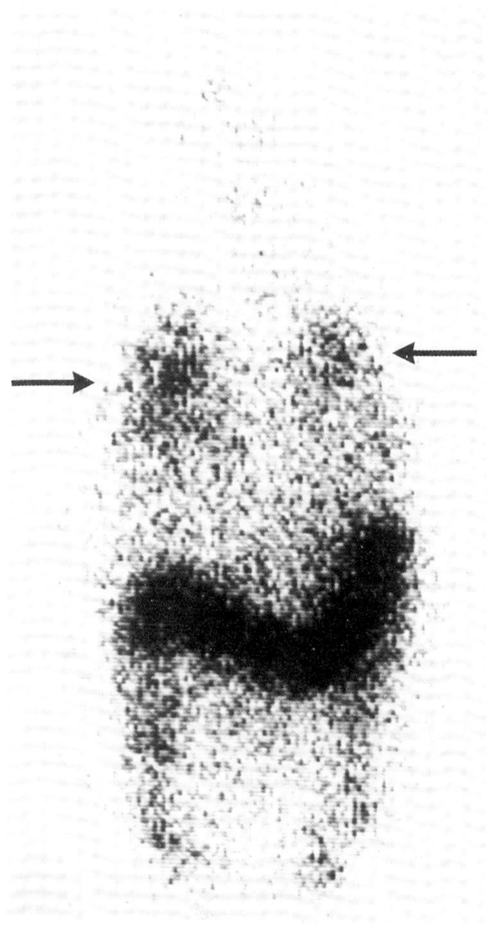

QUESTIONS

Q1. What are the arrowed abnormalities?
Q2. What is the physiological distribution of ^{131}I?
Q3. What types of radiation are emitted by ^{131}I?

ANSWERS

Q1. Activity in lungs due to secondary tumour spread.

Q2. Salivary and lacrimal glands, nasal sinuses, thyroid, stomach, bladder. Bowel activity is due to passage of tracer from stomach.

Q3. Beta radiation, which is useful for therapy, and gamma radiation (367 keV) used for imaging. The gamma radiation means precautions have to be taken following therapeutic use of ^{131}I to minimize radiation dose to others.

TEACHING POINTS

Radioiodine is the optimal radiopharmaceutical to demonstrate recurrence of spread of well differentiated carcinoma of the thyroid. Whole body scans are performed after surgical removal of thyroid and ablation of remaining normal thyroid. Patients have to stop thyroxine replacement therapy for one month to produce a thyroid stimulating hormone (TSH) rise in order for radioiodine to demonstrate secondary tumour to be present.

FURTHER READING

Dworkin, H., Meier, D.A. and Kaplan, M. (1995) Advances in the management of patients with thyroid disease. *Semin. Nuc. Med.*, **XXV**, 205–20.

Case 19

HISTORY

A 54-year-old woman with carcinoid syndrome was referred for assessment for radionuclide therapy. An anterior abdominal view is shown, with marked increased liver activity.

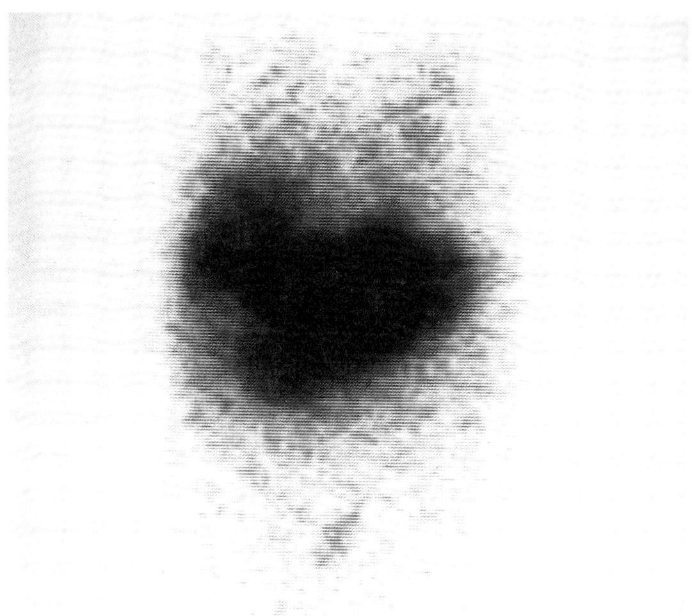

QUESTIONS

Q1. Which radiopharmaceutical was used?
Q2. What therapy would be considered following this test?

ANSWERS

Q1. ^{123}I MIBG.
Q2. ^{131}I MIBG.

TEACHING POINTS

Metaiodobenzyl guanidine (MIBG) is taken up by neuro-endocrine tumour, e.g. carcinoid, phaeochromocytoma, medullary carcinoma of the thyroid and neuroblastoma. ^{123}I MIBG is preferred to ^{131}I MIBG for imaging with better photon flux and lower radiation dose. If uptake is demonstrated with ^{123}I MIBG, ^{131}I MIBG can be used for therapy. When using both ^{123}I and ^{131}I MIBG it is important to remember to block uptake into the thyroid to reduce the radiation dose and, in patients having ^{131}I MIBG, to prevent thyroid ablation. ^{111}In octreotide can also detect carcinoid tumours and is probably more sensitive than ^{123}I MIBG. The demonstration of uptake with ^{111}In octreotide may help select patients for 'cold' octreotide therapy as somatostatin receptors are demonstrated to be present, but this role of octreotide scanning requires further evaluation.

Case 20

HISTORY

A 10-year-old child presented with a history of intermittent rectal bleeding. No cause was apparent. A pertechnetate scan of the abdomen was requested. One- and five-minute views post-injection are shown.

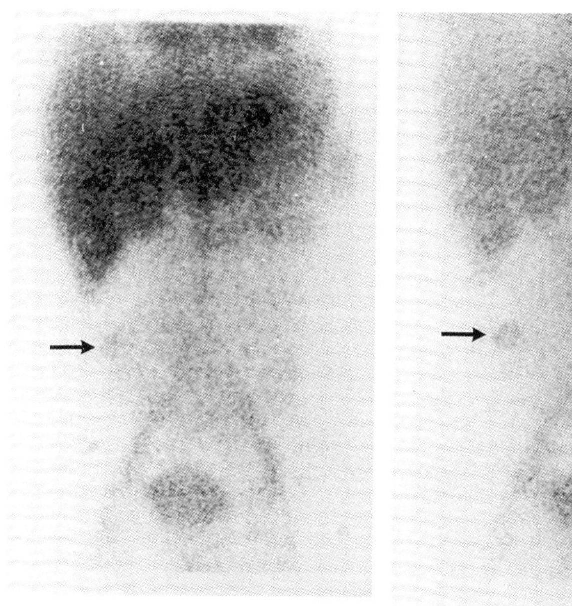

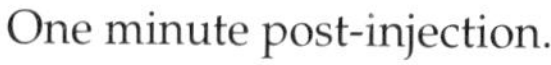

One minute post-injection.

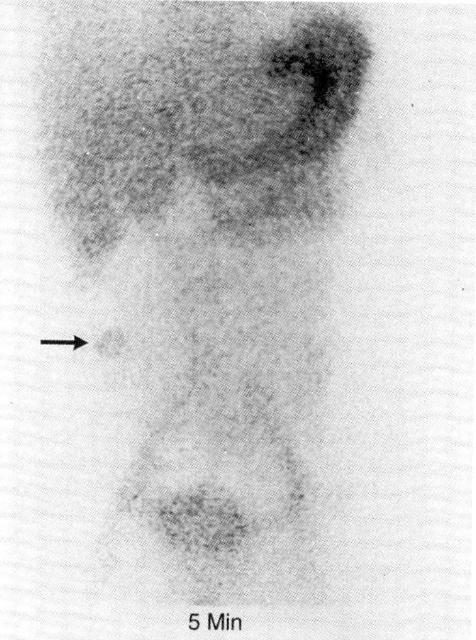

Five minutes post-injection.

QUESTIONS

Q1. What is the normal physiological distribution of pertechnetate?
Q2. What is the arrowed abnormality?

ANSWERS

Q1. The normal uptake of pertechnetate is the same as radioiodine, see Case 18.

Q2. The arrowed abnormality is uptake in a Meckel's diverticulum.

TEACHING POINTS

Pertechnetate is normally taken up by the gastric parietal cells of the stomach. If a Meckel's diverticulum is present, activity should appear at the same time as the stomach. This is usually noted in the right iliac fossa.

FURTHER READING

Berquist, T.H., Nolan, N.G., Stephens, D.H. and Carlnon, H.C. (1976) Specificity of ^{99m}Tc pertechnetate in scinitigraphy diagnosis of Meckel's diverticulum: review of 100 cases. *J. Nuc. Med.*, **17**, 465–9.

Case 21

HISTORY

A 10-year-old boy presents with a two-year history of back pain. Symptoms are worse at night. X-ray of the spine was normal. A two-phase (blood pool and delayed image) bone scan with SPECT was performed.

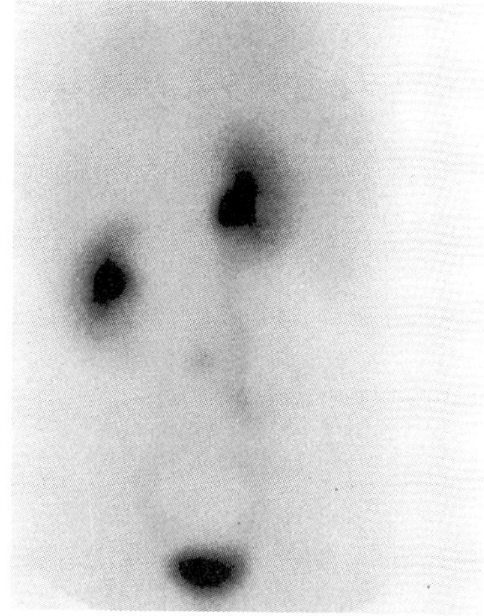

Posterior spine (blood pool).

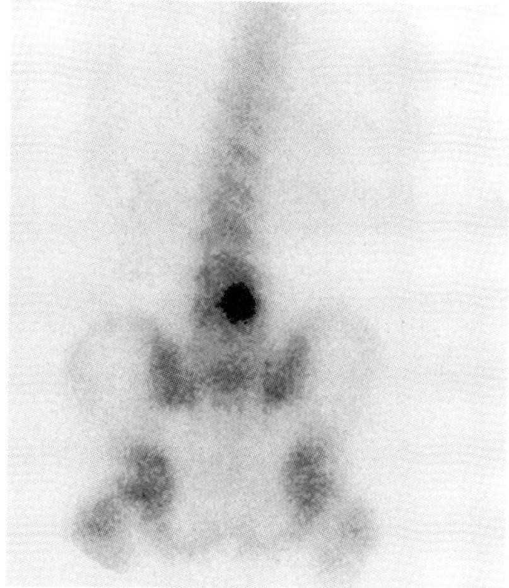

Posterior lumbar spine (delayed image).

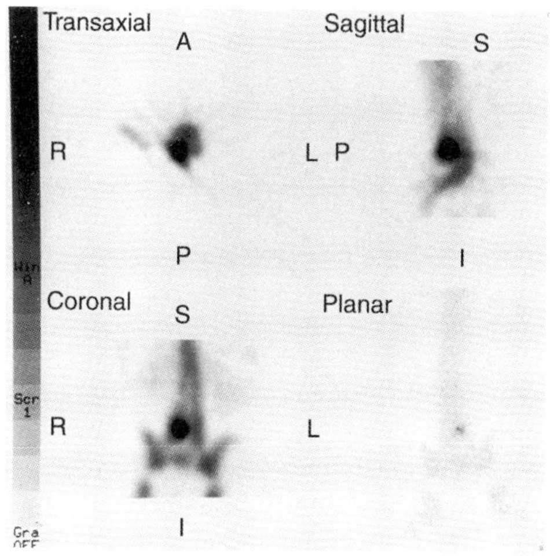

Lumbar spine (SPECT).

QUESTIONS

Q1. Where is the abnormality?
Q2. What is the probable diagnosis?
Q3. What further tests are appropriate?

ANSWERS

Q1. Right side L4/5, which is shown to lie in the posterior elements posterolaterally on SPECT.
Q2. Osteoid osteoma.
Q3. A CT will enable precise anatomical localization of the abnormality. In this patient it was located in the superior facet.

TEACHING POINT

Osteoid osteoma is very easily detected on bone scan, with appearances usually of an intense focus of increased activity associated with increased blood pool. A single focus of activity in the facet joint or pars interarticularis is highly unlikely to be malignant, and SPECT is helpful to reduce the diagnostic possibilities. X-rays are often negative in osteoid osteoma, and the bone scan is the investigation of choice.

FURTHER READING

Ryan, P.J. (1994) SPECT bone scanning in the 1990s. *App. Radiol.*, **May**, 30–46.

Case 22

HISTORY

A 52-year-old woman had pains and swelling of her left hand two months following a Colles' fracture. A two-phase bone scan of the hands and wrists was performed.

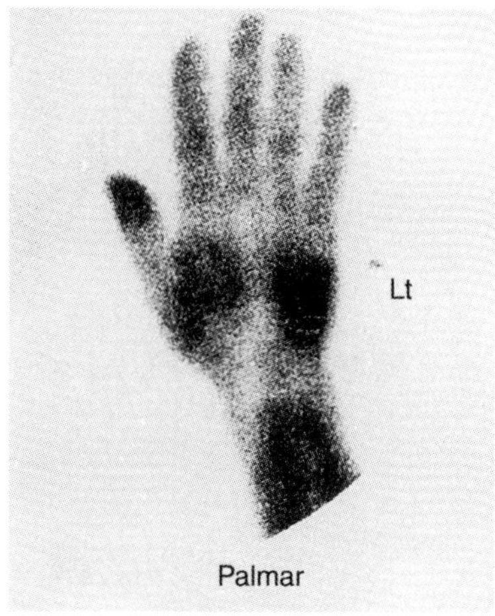

Blood pool.

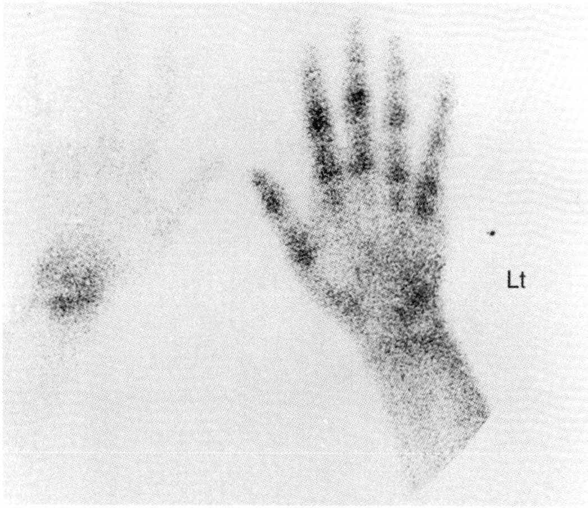

Delayed image.

QUESTIONS

Q1. What is the scan abnormality?
Q2. What is the diagnosis?

ANSWERS

Q1. Increased blood pool and delayed activity throughout the left hand and wrist.

Q2. Reflex sympathetic dystrophy syndrome (RSDS).

TEACHING POINT

RSDS classically shows increased uptake in all the bones of the distal part of an affected limb with striking periarticular uptake. There is usually, but not always, increased activity on blood flow and blood pool images. The bone scan is more sensitive and accurate than x-rays for diagnosis, and has particular advantages in the early stages before the development of radiological osteopenia.

FURTHER READING

Tadorovic-Tiranic, M., Obradovic, V., Han, R. *et al.* (1995) Diagnostic approach to reflex sympathetic dystrophy after fracture: radiography or bone scintigraphy. *Eur. J. Nuc. Med.*, **25**, 1187–93.

Case 23

HISTORY

A 45-year-old woman presented with hypertension and was found to have elevated levels of vanillylmandelic acid (VMA) in her urine. A ^{123}I MIBG study was performed. A posterior view of the chest and abdomen is shown.

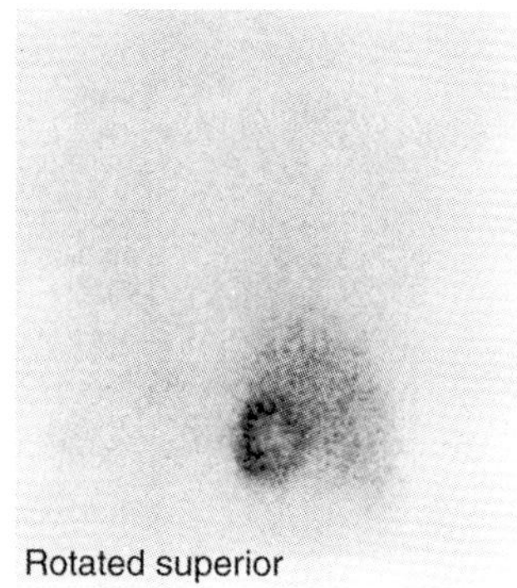

^{123}I MIBG study, posterior view.

QUESTIONS

Q1. What does the study show?
Q2. What is the probable diagnosis?

ANSWERS

Q1. A large area of increased activity in the region of the right adrenal gland.
Q2. Right-sided phaeochromocytoma.

TEACHING POINT

Iodine-labelled MIBG is taken up into 80–90% of phaeochromocytomas. Imaging is useful to confirm the diagnosis, to identify the site of tumour, to locate tumour if in an ectopic position, and to identify metastases (10% of phaeochromocytomas are malignant). If metastases are present, then ^{131}I MIBG can be used for therapy.

Case 24

HISTORY

A 24-year-old man is known to have Crohn's disease. He has continued to experience abdominal pain despite the use of steroids and normal blood parameters. A ^{111}In white cell scan was performed to assess disease activity. Four- and 24-hour anterior and posterior views of the chest and abdomen are displayed.

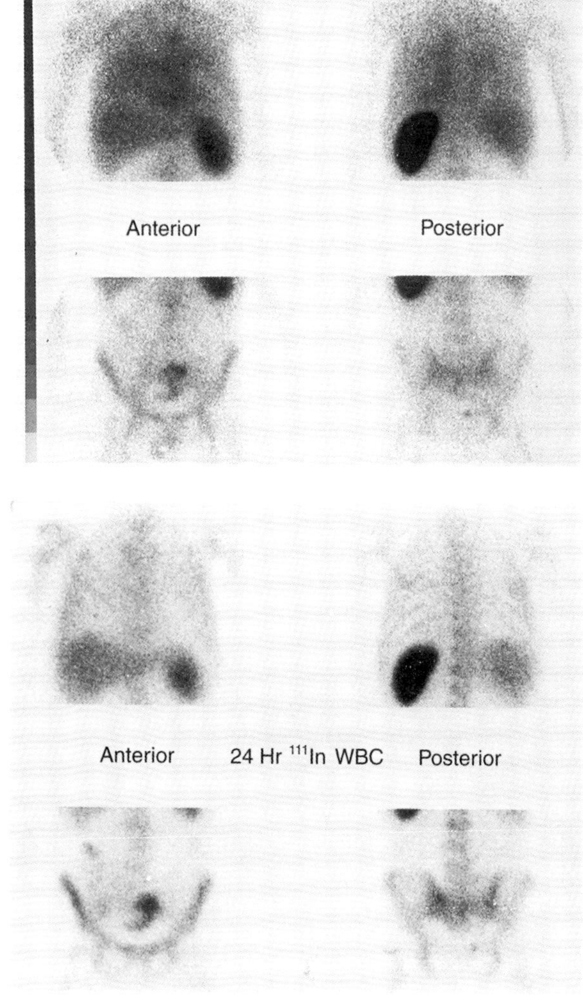

QUESTIONS

Q1. What is the abnormality in the abdomen?
Q2. What does this imply?

ANSWERS

Q1. Bowel activity is visualized in the right abdomen and pelvis, with some change in position of activity on the four- and 24-hour views.
Q2. Continued active Crohn's disease.

TEACHING POINTS

Uptake of indium white cells in bowel is always abnormal. Note that there is no renal excretion of indium compared with Tc white cells, so pelvic activity cannot be confused with the bladder. Imaging of Crohn's disease can be with either technetium- or indium-labelled white cells. Technetium imaging is performed at approximately one and three hours after injection, and this enables the study to be completed in one day. There is also the advantage of a higher count rate. The disadvantages of technetium white cells are non-specific bowel activity after four hours, which makes interpretation of delayed images difficult. Also, free pertechnetate will be excreted via the kidneys; renal and bladder activity can produce confusion. The overall sensitivity of labelled white cells for inflammatory bowel disease is virtually 100%, with specificity of 90–95% depending on the clinical situation.

FURTHER READING

Malcolm, P.N., Bearcroft, C.P., Pratt, P.G. *et al.* (1994) Technetium 99m hexamethyl propyleneamineoxine labelled leucocyte scanning in the initial diagnosis of Crohn's disease. *Br. J. Radiol.*, **67**, 964–8.

Case 25

HISTORY

A 35-year-old man presented with unexplained hypokalaemia. A nuclear medicine scan was performed.

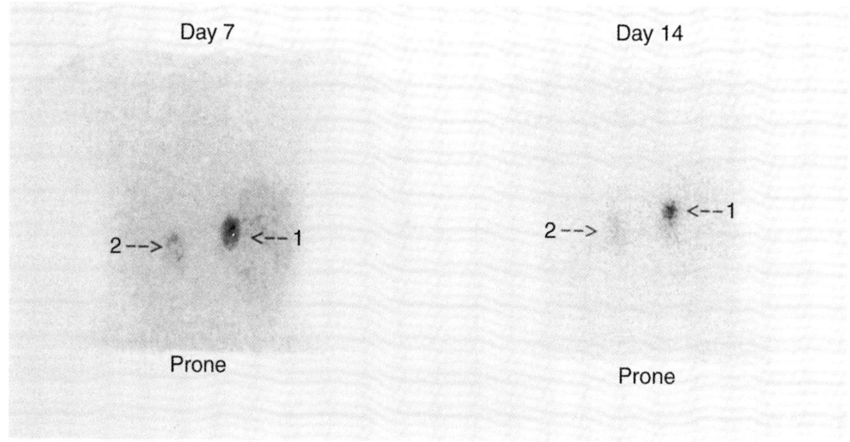

Anterior view of abdomen.

QUESTIONS

Q1. What type of scan was performed?
Q2. What does the scan show?
Q3. What are the implications?

ANSWERS

Q1. A selenocholesterol study.
Q2. More activity at site of left adrenal gland (1) than right (2).
Q3. A Conn's syndrome adenoma.

TEACHING POINTS

Q1. Selenocholesterol scanning of the adrenal is useful for the detection of Conn's syndrome adenomas, the evaluation of Cushing's syndrome, and unexplained adrenal masses.
Q2. In some centres, dexamethasone is given before and for some days after injection to suppress uptake into the adrenal glands. Uptake within five days of injection is then considered abnormal and indicates an adenoma.

Case 26

HISTORY

A 36-year-old man with known carcinoma of the lung presented with left rib pain. A bone scan was performed.

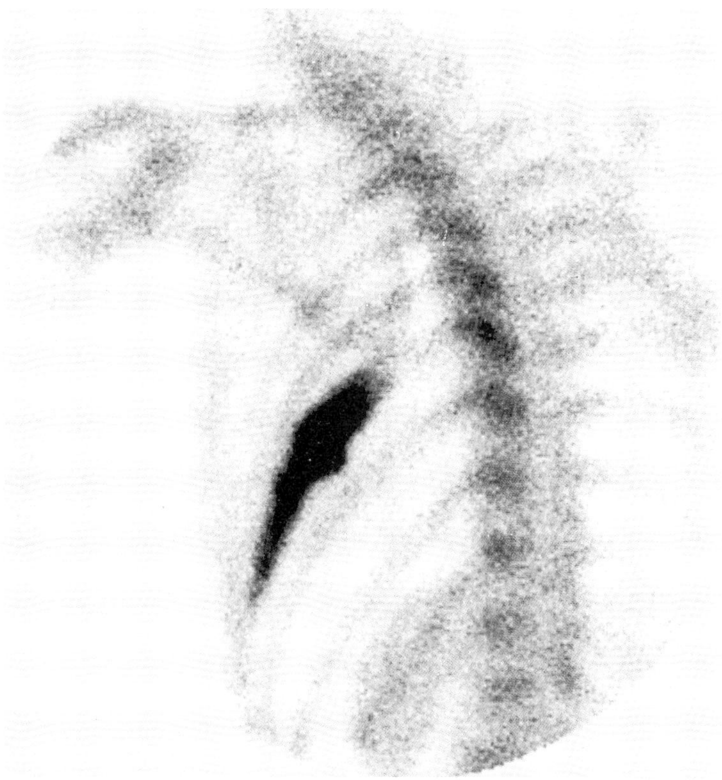

Left posterior oblique view of chest.

QUESTIONS

Q1. What abnormality is shown?
Q2. What does this suggest?

ANSWERS

Q1. Marked increased activity extending along the left 8th rib posteriorly.
Q2. Metastatic disease in the rib.

TEACHING POINT

Activity extending along a rib is virtually diagnostic of malignancy. The major differential diagnosis of rib uptake is a fracture, where activity is usually focal and commonly involves multiple ribs in a linear pattern.

Case 27

HISTORY

A 57-year-old woman had chronic back pain. She was tender at L4/5. Planar and tomographic bone scan images of the lumbar spine were performed. Abnormal uptake is identified (arrowed) on both images.

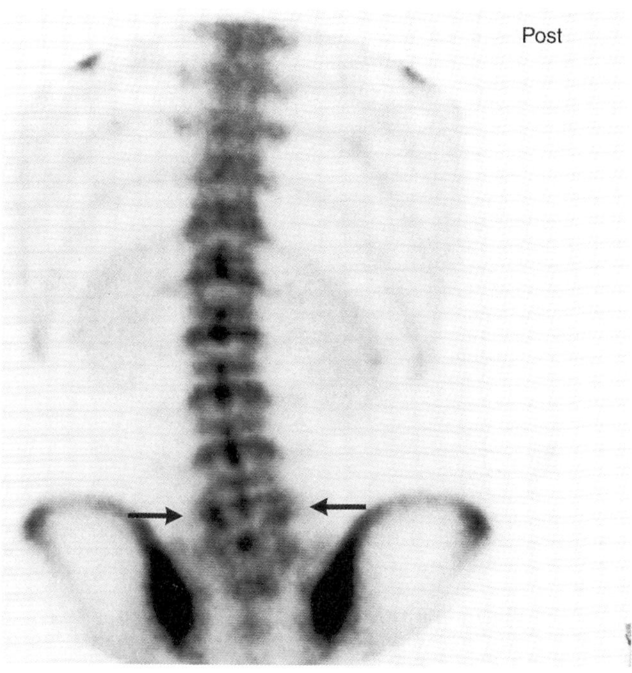

Posterior lumbar spine.

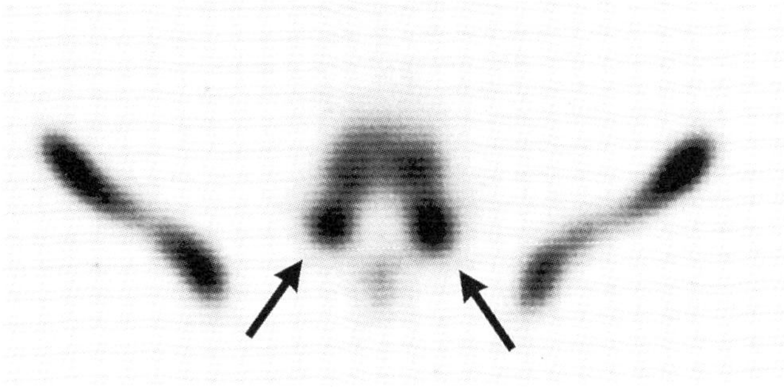

Transaxial SPECT imaging at L4/5.

QUESTIONS

Q1. What is the site of the abnormality?
Q2. What is the probable diagnosis?

ANSWERS

Q1. Increased activity bilaterally in the posterior elements, posterolaterally.

Q2. The most common cause for this is facet joint osteoarthritis. Bilateral fractures of the pars interarticularis can show the same abnormality on transaxial SPECT, but x-rays or a CT scan should be able to identify these.

TEACHING POINT

Active facet joint activity is identified in approximately 30% of patients with chronic back pain, and predicts response to joint injection therapy.

FURTHER READING

Ryan, P.J., Gibson, T. and Fogelman, I. (1992) The identification of spinal pathology in chronic low back pain using single photon emission computed tomography. *Nuc. Med. Comm.*, **13**, 497–502.

Case 28

HISTORY

A 26-year-old woman had persistent pain in her left knee following a twisting injury. A two-phase bone scan of the knee with SPECT was performed.

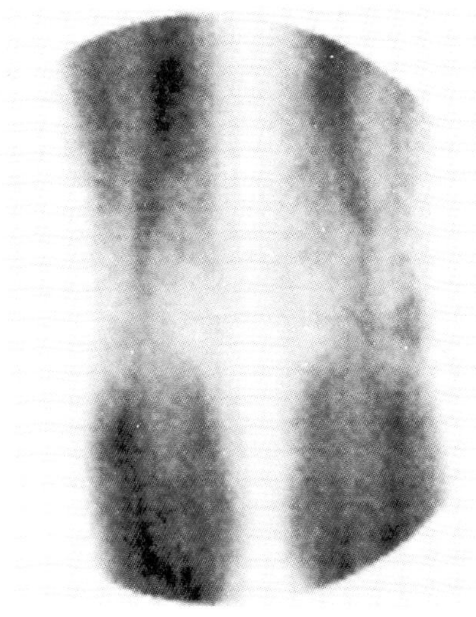

Posterior blood pool.

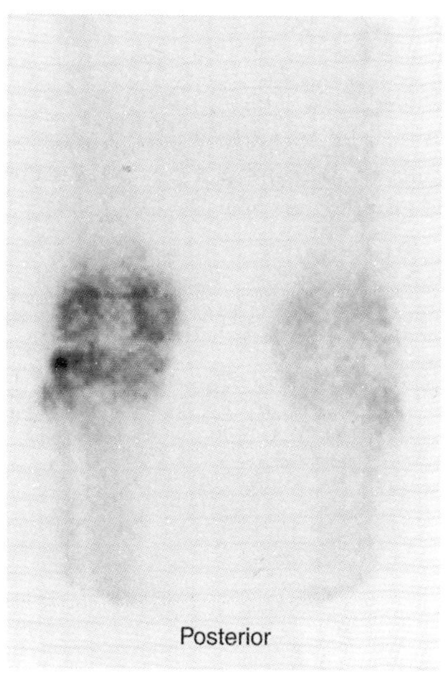

Static posterior knees.

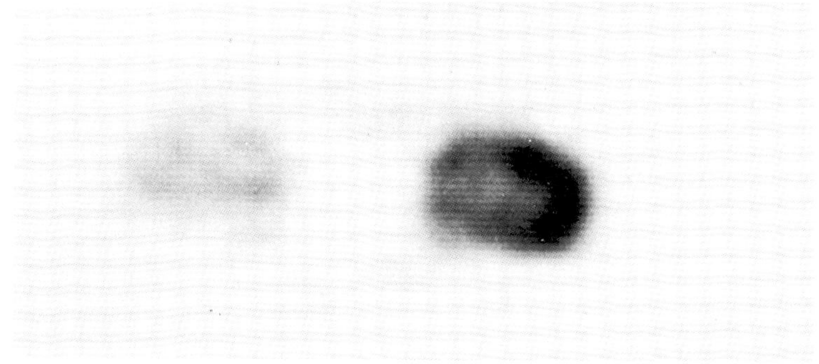

Transaxial SPECT knees.

QUESTIONS

Q1. What abnormalities are shown?
Q2. What is the diagnosis?

ANSWERS

Q1. Increased activity on blood pool and static images lateral left knee with activity localized to the tibial plateau. The transaxial SPECT image shows activity throughout the left lateral tibial plateau in the form of a crescent.

Q2. Tear left lateral meniscus.

TEACHING POINT

Bone SPECT has a high sensitivity and specificity for meniscal tears with the classic features being increased blood pool, activity in tibial plateau on planar study and a peripheral crescent of increased activity in the tibial plateau on transaxial SPECT with adjacent focal posterior femoral condyle activity.

FURTHER READING

Ryan, P.J., Taylor, M., Grevitt, M. *et al.* (1993) Bone single photon emission tomography in recent meniscal tears; an assessment of diagnostic criteria. *Eur. J. Nuc. Med.*, **20**, 703–7.

Case 29

HISTORY

A 63-year-old man presented with haemoptysis and a mass was visualized on chest x-ray. Bronchoscopy and biopsy revealed squamous carcinoma of the bronchus. A staging PET study was performed.

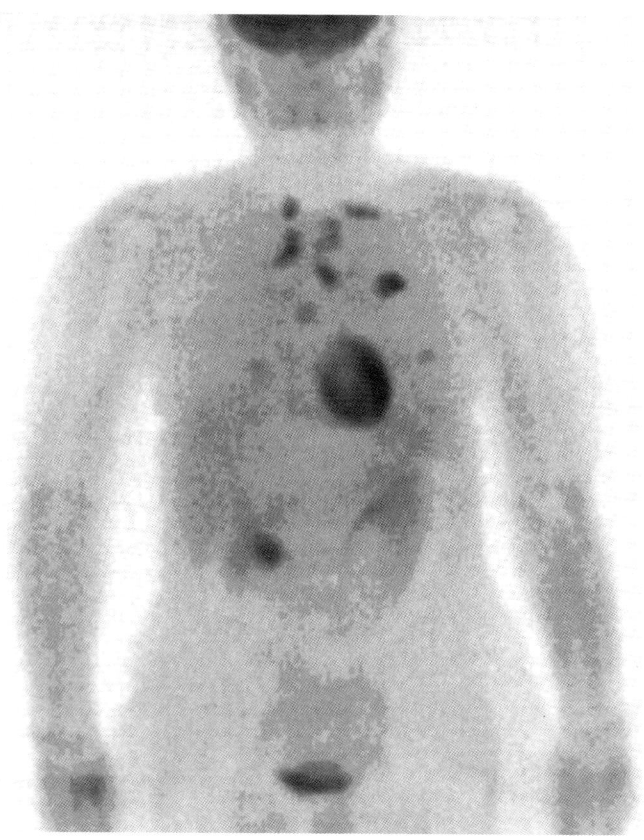

Whole body PET scan.

QUESTIONS

Q1. What agent was used?
Q2. What does the scan show?
Q3. How does it help in this condition?

ANSWERS

Q1. ^{18} Fluorine deoxyglucose (FDG).

Q2. Normal uptake in brain, heart and renal tract and bowel. There are multiple foci of abnormal uptake in chest representing metastases.

Q3. FDG is taken up preferentially by tumours due to hypermetabolism of glucose compared to normal cells and becomes fixed in the cell. Although it is taken up like glucose it is not metabolized beyond glucose 6-phosphate. FDG PET studies show a very high sensitivity for many malignant tumours and will detect metastases not visualized on conventional imaging (CT and MRI).

TEACHING POINTS

The role of FDG studies in staging cancers has yet to be established but for many cancer tumours such as carcinoma of the breast, lung and lymphoma it appears to be superior to established methods. It has the advantage of being able to evaluate the whole body in one single study.

Case 30

HISTORY

A patient presented wth hypercalcaemia, with an elevated serum parathyroid hormone level. A diagnosis of primary hyperparathyroidism was made. A parathyroid pertechnetate–thallium subtraction scan was performed to localize a possible adenoma.

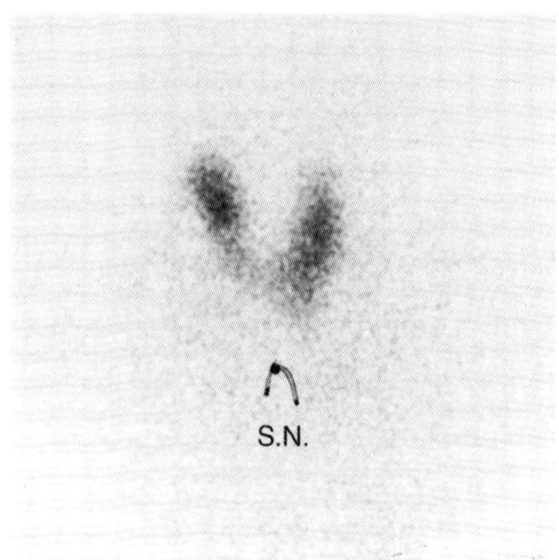

Pertechnetate scan of the neck.

Thallium scan of the neck.

Subtraction image.

QUESTIONS

Q1. What abnormalities are shown?
Q2. What is the diagnosis?

ANSWERS

Q1. There is focal increased activity on the thallium scan in the left lower pole, which is more clearly apparent after subtraction of pertechnetate activity.

Q2. Parathyroid adenoma left lower pole.

TEACHING POINT

A parathyroid subtraction scan is very useful to localize an adenoma in primary hyperparathyroidism prior to surgery. This reduces surgery time and makes the operation simpler. The principle of the scan is that a parathyroid adenoma will take up thallium but not pertechnetate. The study is difficult to interpret where the thyroid scan is abnormal. An important pitfall is a multinodular goitre on pertechnetate imaging, which can give false-positive results due to thallium uptake in non-functioning nodules. Thyroid carcinoma as in the previous case could produce confusion, although thallium uptake should then be at the site of a cold nodule on pertechnetate scan. 2-Methoxyisobutyl isonitrile (MIBI) can be used as an alternative to thallium, and is the preferred agent in some centres.

Case 31

HISTORY

A 35-year-old man presented with atrial fibrillation and cardiac failure. He had been previously well and had no history of rheumatic fever or excessive alcohol consumption. A ^{111}In myoscint study was performed to investigate further. Images are shown 48 hours after injection.

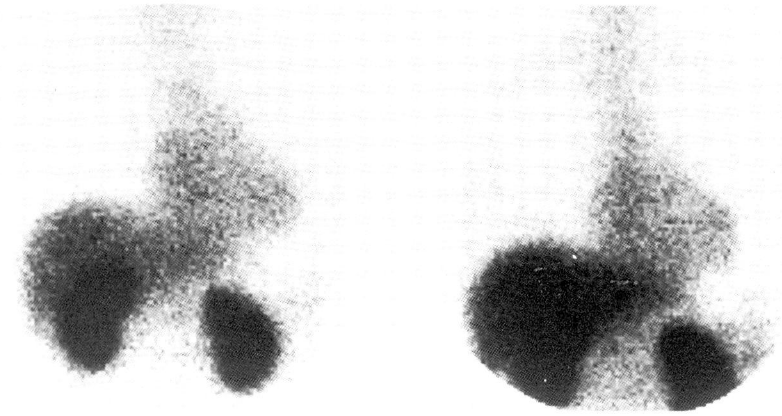

Myoscint study.

QUESTIONS

Q1. What does the study show?
Q2. What is the diagnosis?

ANSWERS

Q1. Uniform increased uptake of tracer throughout the myocardium.
Q2. Myocarditis.

TEACHING POINT

Myoscint is a murine Fab fragment against myosin coupled with DTPA (diethylenetriamine penta-acetic acid) and labelled with ^{111}In. The antibody binds to intracellular myosin expressed as a result of disease. Imaging is performed at 24 and 48 hours after injection. In a normal heart, there is some increased blood pool at 24 hours, which fades by 48 hours. Retention at 48 hours is abnormal. The main differential diagnosis from myocarditis is myocardial infarction, where uptake is more focal.

FURTHER READING

Carrio, J., Berna, L., Ballester, M. *et al*. (1988) Indium 111 antimyosin scintigraphy to assess myocardial damage in patients with suspected myocarditis and cardiac rejection. *J. Nuc. Med.*, **29**, 1893–1900.

Case 32

HISTORY

A 55-year-old man with ankylosing spondylosis developed a painful right heel. He was tender medial and distal to the heel pad. A two-phase bone scan of the feet was performed.

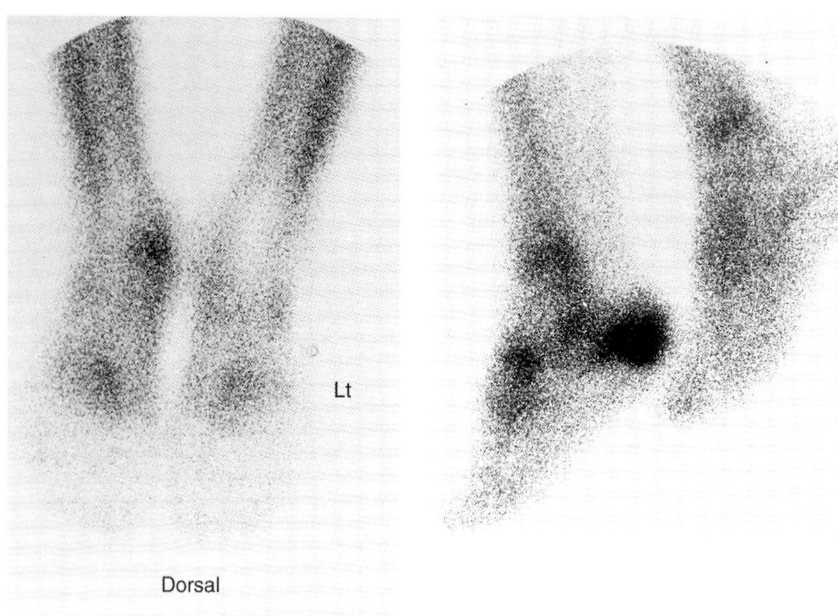

Blood pool. Delayed image (right medial left lateral).

QUESTIONS

Q1. Describe the abnormality?
Q2. What is the diagnosis?

ANSWERS

Q1. Increased tracer uptake right heel, medially and inferiorly, with associated increased blood pool.

Q2. Plantar fasciitis.

Case 33

HISTORY

A 60-year-old woman with known carcinoma of the lung was investigated with a whole body bone scan for the presence of skeletal metastases.

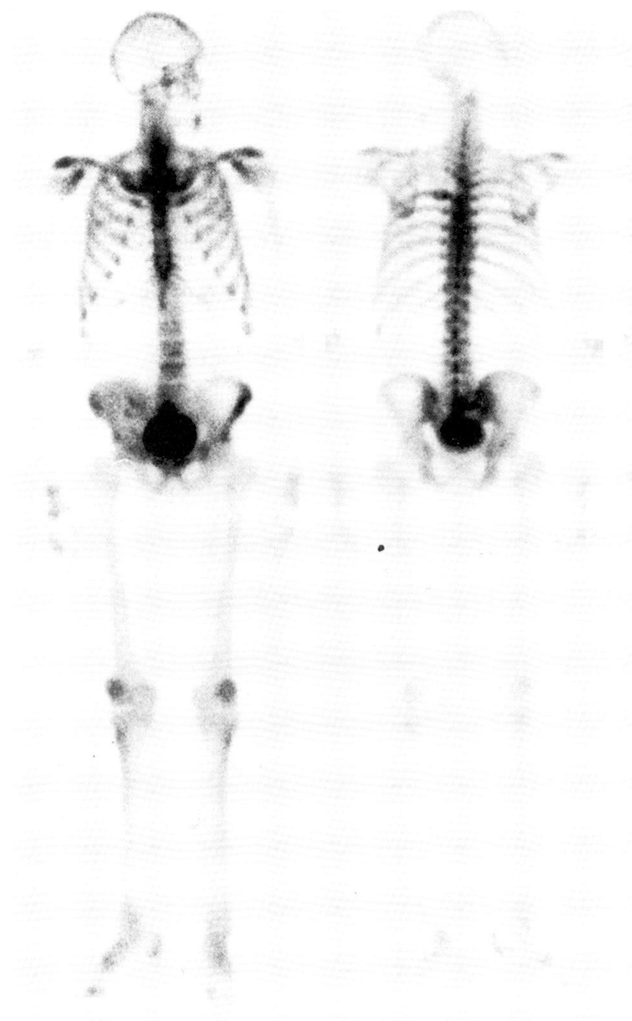

QUESTIONS

Q1. What abnormalities are demonstrated?
Q2. What does this imply?

ANSWERS

Q1. There is increased tracer uptake left 6th rib posteriorly and right hemipelvis. There is a photon deficient area surrounded by a rim of increased activity right sacroiliac joint. The left 5th rib posteriorly is not visualized. Kidneys are not seen.

Q2. Skeletal metastases. Absent left 5th rib posteriorly is due to surgical removal. Non-visualization of kidneys implies metastatic superscan.

TEACHING POINT

Skeletal metastases may show reduced as well as increased uptake where there is aggressive lytic and destructive disease, although this is far less common.

Case 34

HISTORY

A 64-year-old man presented with bleeding per rectum. Upper gastrointestinal endoscopy was normal. Bleeding failed to settle with conservative measures, and surgery was contemplated. A ^{99m}Tc labelled red blood cell study was performed to aid localization of the bleeding site. Anterior abdominal images are displayed immediately after injection and 10 minutes and 60 minutes later.

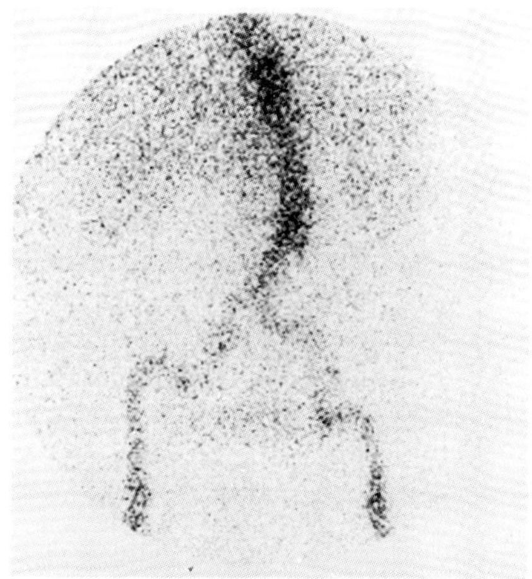

Immediately post-injection (0 minutes).

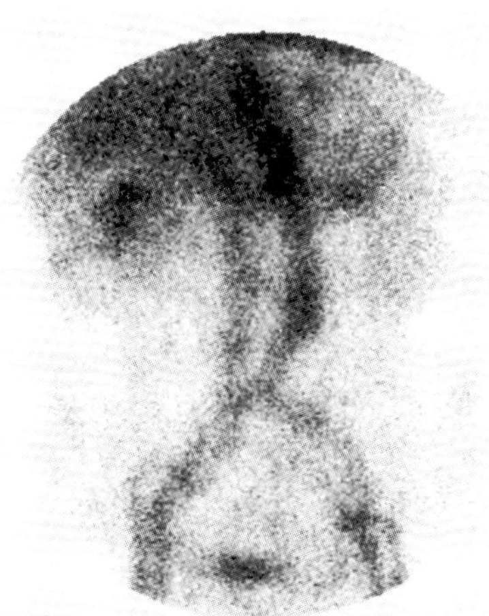

10 minutes post-injection.

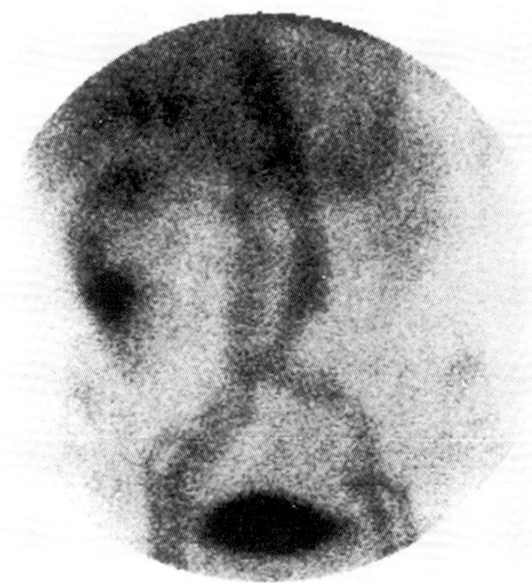

60 minutes post-injection.

QUESTIONS

Q1. What is the abnormality?

Q2. What information does the study provide regarding the site of bleeding?

ANSWERS

Q1. At 10 minutes there is activity in a loop of bowel extending across the upper abdomen. This is more clearly identified at 60 minutes and would appear to be in the upper right ascending colon and transverse colon.

Q2. Bowel uptake on a red blood cell study is abnormal. No activity is identified proximal to the mid-right colon, which is the probable site of bleeding.

TEACHING POINTS

1. A red blood cell study will only be positive if the patient is bleeding during the length of the study.
2. A red blood cell study may be positive when angiographically negative, and is often used as the investigation of choice following endoscopy.
3. A red blood cell study will often localize the probable site of bleeding, although this will always be at or proximal to the initial abnormal activity seen.

Case 35

HISTORY

A 25-year-old intravenous drug abuser presented with a dry cough, dyspnoea and weight loss. He had a normal x-ray and no sputum could be obtained. A ^{111}In polyclonal immunoglobulin study was obtained. Anterior images of the chest are displayed at four hours and 24 hours after injection.

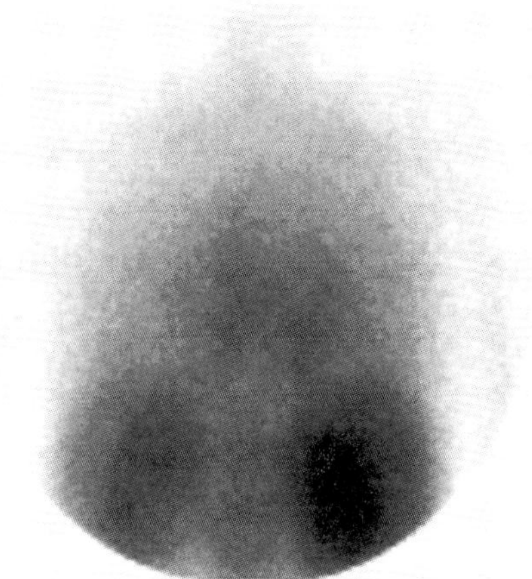

Four hours post-injection.

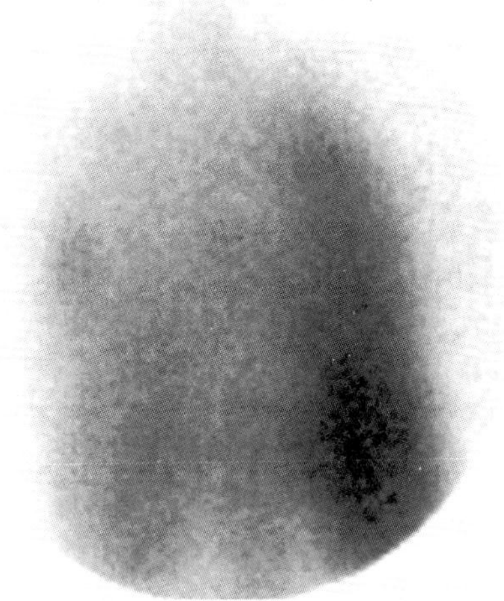

24 hours post-injection.

QUESTIONS

Q1. Describe the abnormalities.
Q2. What do they imply?

ANSWERS

Q1. There is increased activity left lung and right lower lobe in a diffuse pattern. There is probable focal uptake right mid-zone.

Q2. Diffuse increased activity suggests *Pneumocystis carinii* pneumonia (PCP), which was confirmed in this patient. Focal uptake right mid-zone implies probable concurrent bacterial infection.

TEACHING POINT

Diffuse uptake on gallium scan or polyclonal immunoglobulin (HIG) images suggested a diagnosis of PCP. HIG imaging is quicker and easier to perform, but there is less experience with this agent than with gallium. In patients with suspected PCP but where diagnosis cannot be made on chest x-ray and sputum tests, a nuclear medicine study can be helpful. Some centres also perform a ^{99m}Tc DTPA permeability study in which a biphasic washout curve implies PCP.

FURTHER READING

Doherty, M. and Nunan, T.O. (1993) Nuclear medicine and AIDS. *Nuc. Med. Comm.*, **14**, 830–48.

Case 36

HISTORY

A 56-year-old woman with known breast carcinoma presented with general poor health and weight loss. An enlarged liver was found on examination. Ultrasound was technically difficult and an isotope liver scan was performed. Anterior and posterior views are displayed.

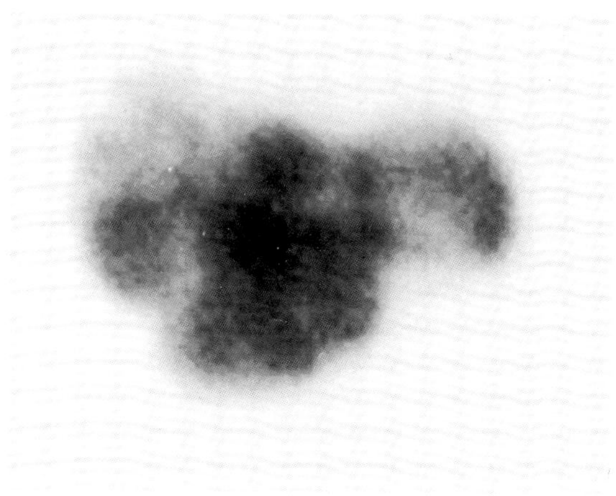

Anterior view.

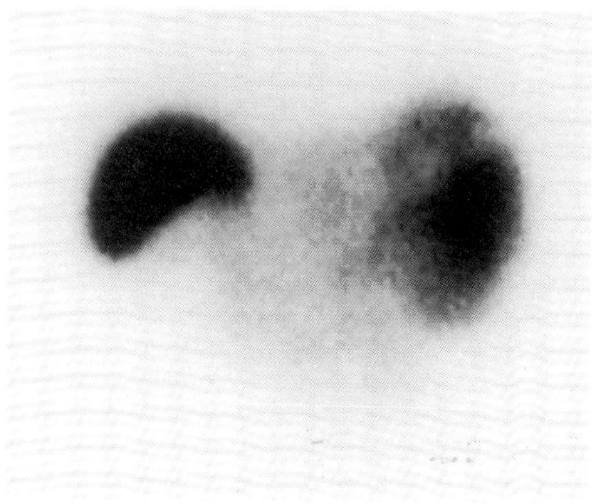

Posterior view.

QUESTIONS

Q1. What is abnormal on this study?
Q2. What is the diagnosis?

ANSWERS

Q1. Hepatomegaly with multiple focal photon deficient areas throughout the liver.

Q2. Liver metastases.

TEACHING POINT

The isotope liver scan is usually performed as a screening test for liver metastases only when ultrasound is not successful. It remains a sensitive test for liver metastases, however and is particularly valuable for assessment of disease progression/regression following therapy, as hard copy images can be easily compared.

Case 37

HISTORY

A 62-year-old man fractured his right femur. The fracture was treated by manipulation and fixation in plaster.

A year after the fracture, he continued to suffer pain. A two-phase bone scan was performed.

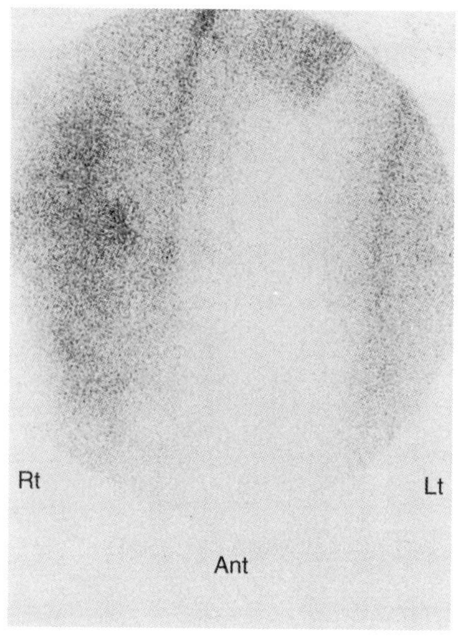

Blood pool.

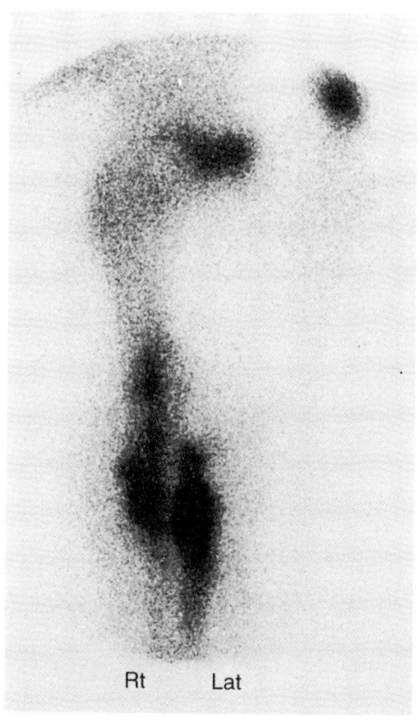

Anterior delayed view.

QUESTIONS

Q1. What is the pattern of abnormal activity?
Q2. How does this help the orthopaedic surgeon?

ANSWERS

Q1. There is increased tracer activity with increased blood pool at both sides of the fracture.

Q2. The diagnosis is non-union. Where there is reduced activity between the bone ends, there is little chance of union without further surgery.

FURTHER READING

Gunalp, B., Ozguven, M., Oztrul, E. and Bayhan, H. (1992) Role of bone scanning in the management of non-united fractures: a clinical study. *Eur. J. Nuc. Med.*, **19**, 845–7.

Case 38

HISTORY

A six-month-old baby was being investigated for bruising. A radiological survey failed to demonstrate a fracture. A bone scan was therefore performed.

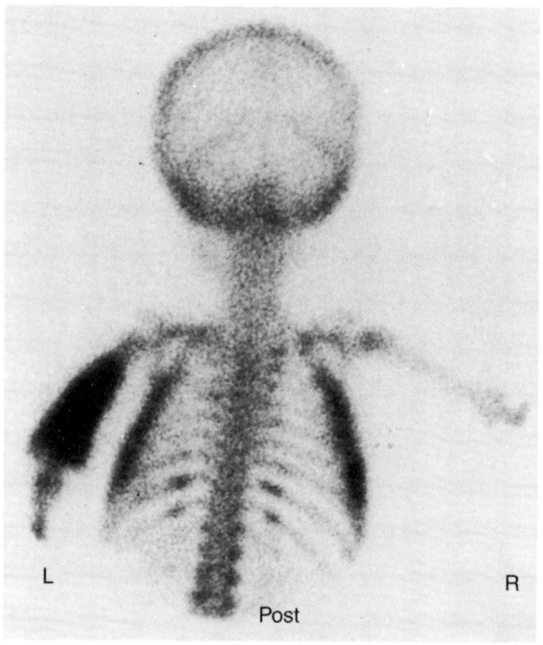

Posterior chest.

QUESTIONS

Q1. What abnormalities are shown?
Q2. What do they imply?

ANSWERS

Q1. Increased tracer activity throughout most of the left humerus. Bilateral focal rib activity 10th, 11th ribs posteriorly and multiple ribs laterally.

Q2. Multiple fractures due to non-accidental injury (NAI).

TEACHING POINTS

1. Bone scintigraphy is useful in NAI to detect fractures if x-rays are negative and in all cases to evaluate the whole skeleton.
2. NAI should be suspected when the bone scan identifies random lesions in the skeleton in the absence of a history of trauma.
3. In children in particular, fractures may show diffuse increased uptake rather than focal activity, especially in the first weeks after injury.
4. Skull x-rays should always be performed, as fractures at this site may not be demonstrated on bone scan.

FURTHER READING

Park, H.M., Kevelz, C.B. and Robb, J.A. (1988) Early scintigraphic findings of occult femoral and tibial fractures in infants. *Clin. Nuc. Med.*, **13**, 271–5.

Case 39

HISTORY

A 50-year-old man had long-standing rheumatoid arthritis. He gave a more recent history of vomiting, particularly after meals. Endoscopy failed to show any mucosal lesions. A gastric emptying study was performed. The time activity curves are shown below for liquid and solid phases.

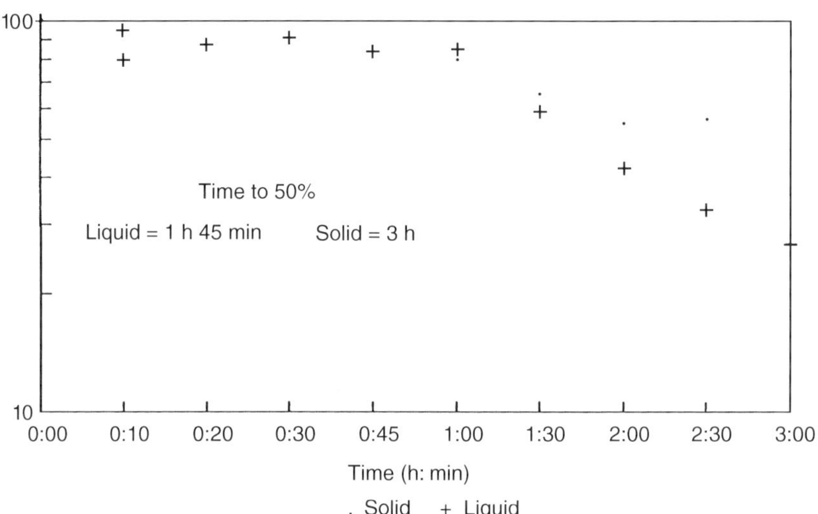

QUESTIONS

Q1. How is a gastric emptying study performed?
Q2. What are the clinical indications?

ANSWERS

Q1. Liquid and solid meals are labelled with radiopharmaceuticals of different gamma energies. This allows simultaneous acquisition of both sets of data. The liquid phase is commonly performed with ^{111}In DTPA, and the solid phase with technetium-labelled bran or a polyester such as Douex. Anterior and posterior images are acquired at five-minute intervals and time activity curves are calculated after correction for radioactive decay. The normal time for half the activity to disappear from the stomach is approximately 60 minutes for the liquid and 90 minutes for the solid phase.

Q2. Clinical symptoms of delayed gastric emptying, which in this case was due to amyloid of the stomach, or rapid gastric emptying, which is sometimes seen after gastric surgery. Images of a patient with normal gastric emptying (liquid phase) are shown below.

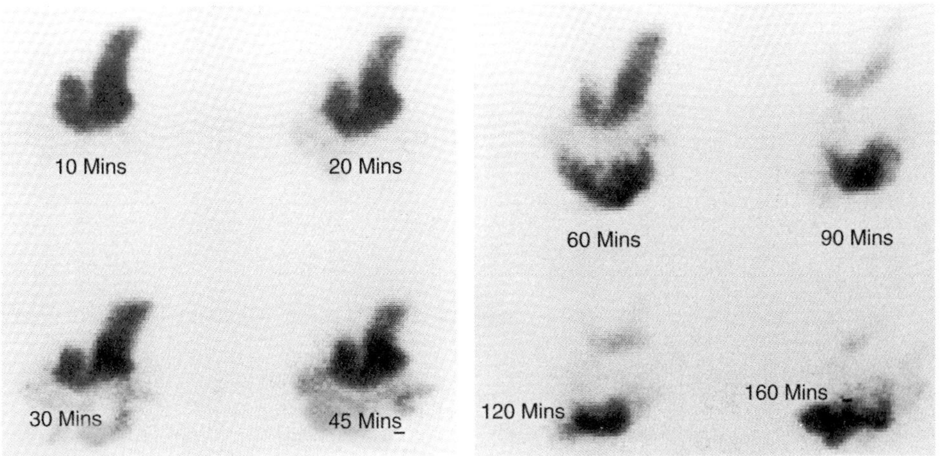

FURTHER READING

Datz, F.L., Christian, P.E., Hutson, W.R. *et al.* (1991) Physiological and pharmacological interventions in radionuclide imaging in the tubular gastrointestinal tract. *Semin. Nuc. Med.*, **XXI**, 140–52.

Case 40

HISTORY

A 34-year-old man had Hodgkin's lymphoma. Treatment was given in the form of chemotherapy and radiotherapy. Possible recurrence was suspected in the chest, although CT findings were non-specific. A gallium scan was performed and the anterior head and chest images are shown below.

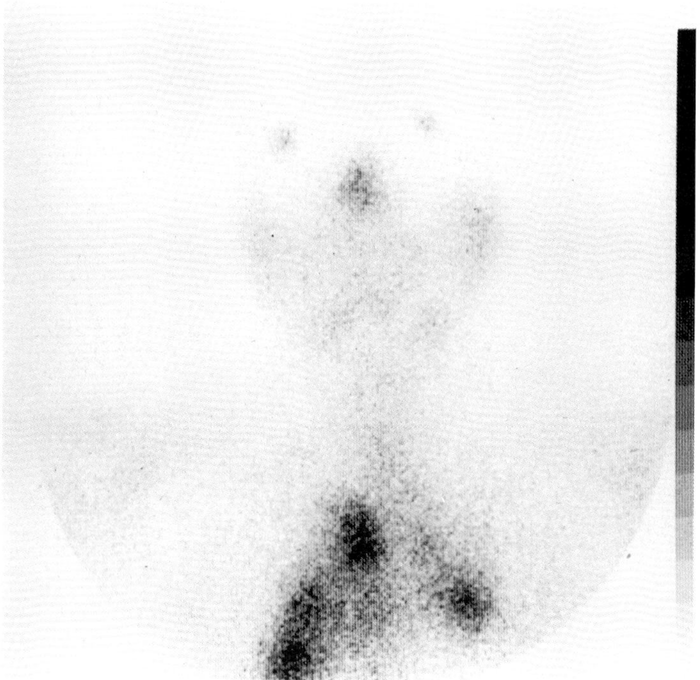

QUESTIONS

Q1. What abnormalities are seen on this study?
Q2. What do these imply?

ANSWERS

Q1. Marked increased uptake in hilar and suprahilar lymph nodes.
Q2. Recurrence of Hodgkin's lymphoma.

TEACHING POINT

Gallium is frequently taken up by lymphoma. Although not commonly used for staging, it is a helpful agent for distinguishing fibrotic change following radiotherapy from tumour recurrence. This is a situation where CT and MRI may be difficult to interpret.

FURTHER READING

Kostalzaglu, L., Heh, S.D.J., Portlock, C. *et al.* (1992) Validation of gallium 67 citrate single photon emission computed tomography in biopsy confirmed residual Hodgkin's disease in the mediastinum *J. Nuc. Med.*, **33**, 345–50.

Case 41

HISTORY

A 30-year-old male presented with weight loss. A gallium scan was performed.

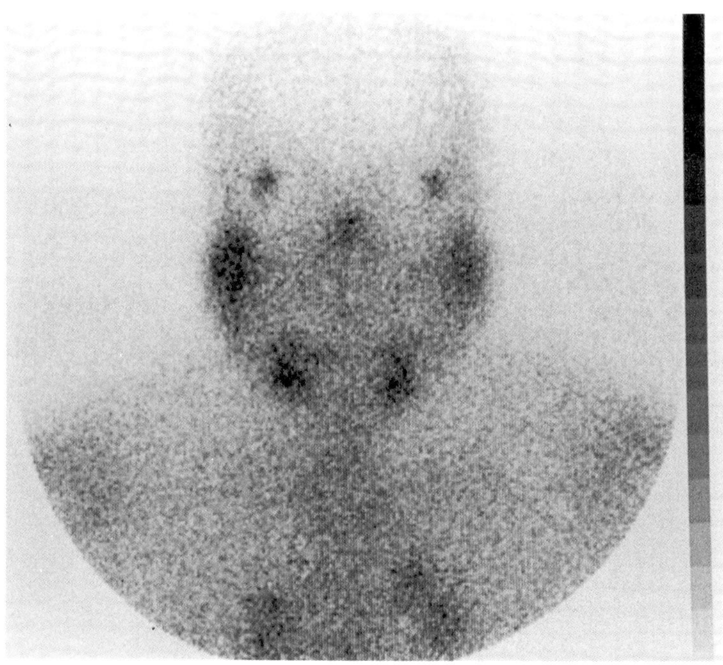

QUESTIONS

Q1. Describe the scan.
Q2. What is the most likely diagnosis?

ANSWERS

Q1. The areas of abnormal uptake are:
- bilateral parotid uptake
- bilateral submandibular uptake
- bilateral lacrimal uptake
- bilateral hilar uptake.

Q2. Sarcoidosis.

TEACHING POINTS

1. Gallium scanning is used in sarcoidosis to:
   - confirm the diagnosis
   - assess disease extent
   - follow up the effects of therapy
   - differentiate fibrosis from active disease.
2. Lacrimal uptake is non-specific, but the diagnosis of sarcoidosis should be suspected when it is pronounced.

Case 42

HISTORY

An 80-year-old diabetic patient presented with pain and swelling in the right foot. Images from a nuclear medicine study are displayed.

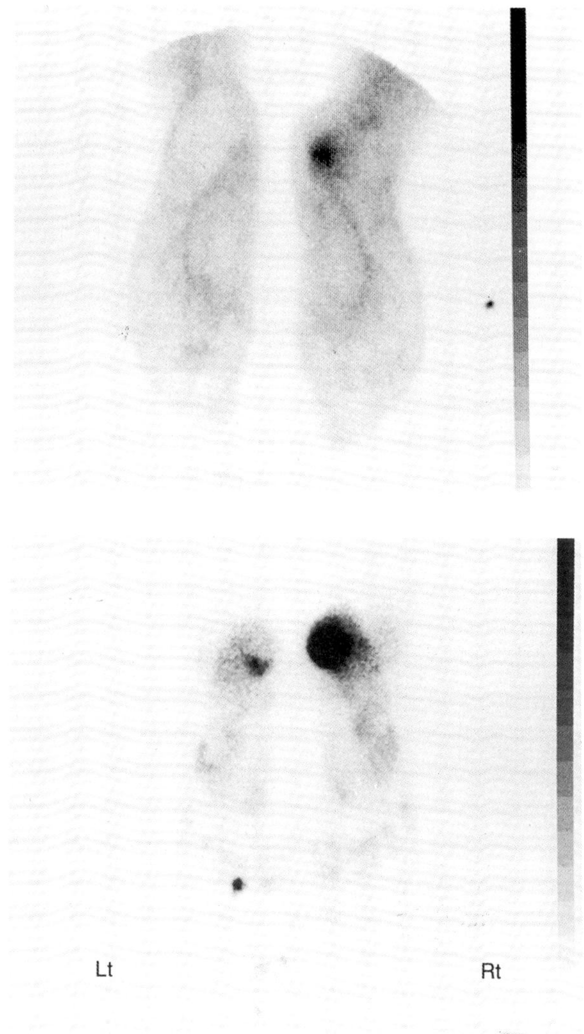

QUESTIONS

Q1. What is this investigation?
Q2. Describe the abnormalities.
Q3. What is your interpretation of the findings?

ANSWERS

Q1. This is a two-phase bone scan.

Q2. There is increased blood pool and increased metabolic activity in the region of the right calcaneus. There is also some increased metabolic activity associated with the metatarsophalangeal joint of the second toe on the left, and also a small focus of increased activity overlying the left calcaneus.

Q3. In view of the clinical history, the increased blood pool, and the metabolic activity associated with the right calcaneus, the most likely diagnosis is infection. A gallium or a white cell scan would confirm this. The appearance of the left foot is likely to be degenerative in origin.

TEACHING POINT

Increased bone scan activity in the feet of diabetic patients may be due to multiple causes, including osteomyelitis and Charcot's arthropathy. A combined bone and white cell scan provides very high sensitivity and specificity for the detection of infection.

FURTHER READING

Kolindou, A., Liv, Y., Ozher, K. *et al.* (1996) InIII WBC imaging of osteomyelitis in patients with underlying bone scan abnormalities. *Clin. Nuc. Med.*, **21**, 183–91.

Case 43

HISTORY

A 70-year-old patient presented with bone pain.

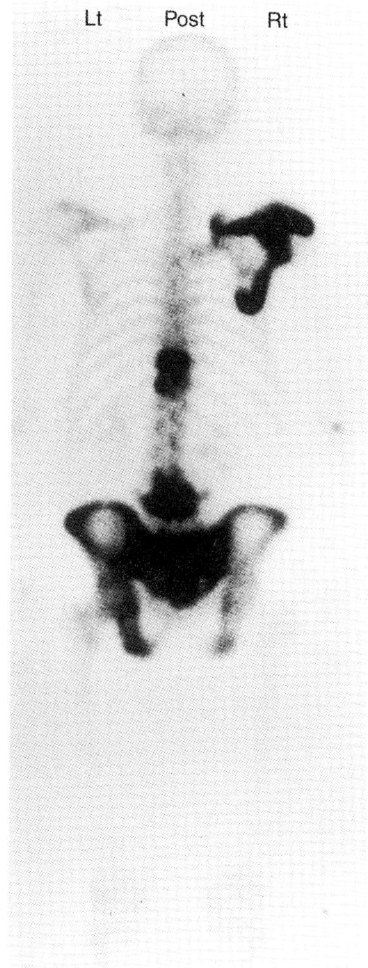

QUESTIONS

Q1. What are the abnormalities on the scan?
Q2. What is the likely diagnosis?
Q3. What two sites are most commonly involved in this disease process?

ANSWERS

Q1. There is intense increased tracer uptake associated with the right scapula, the lower thoracic vertebrae and the L5 vertebra. Increased uptake is also seen in association with the sacrum and both hemipelves.
Q2. Paget's disease.
Q3. The spine and pelvis.

TEACHING POINTS

Typical bone scan features of Paget's disease are:

- intense tracer uptake involving the whole of the affected bone.
- the bone appears expanded
- deformity of the involved bone
- diffuse increased activity
- one or both ends of the long bones involved
- anatomical features are enhanced. For example, the transverse processes of the vertebrae may be seen.

FURTHER READING

Fogelman, I. (1991) Bone scanning in Paget's disease, in *Nuclear Medicine Annual* (ed. L.M. Freeman) Raven Press, New York.

Case 44

HISTORY

A 75-year-old patient developed hip pain following an operation.

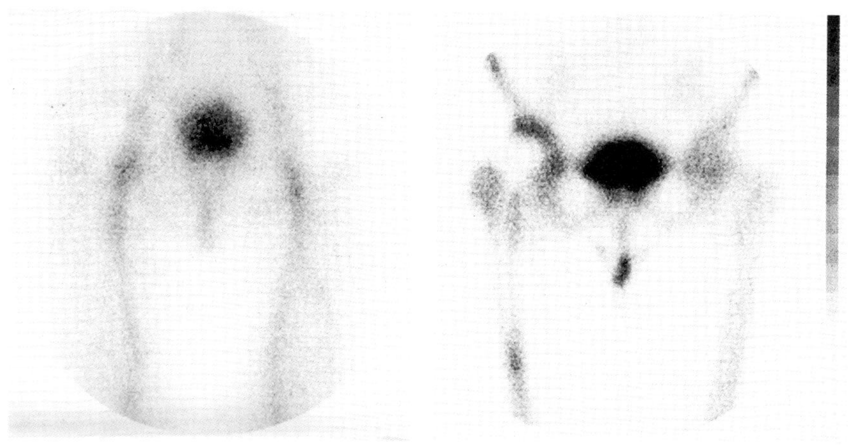

QUESTIONS

Q1. What is this investigation?
Q2. What are the abnormalities on the scan?
Q3. What is the likely diagnosis?

ANSWERS

Q1. This is a two-phase bone scan.

Q2. There has been a right total hip replacement. There is an area of increased activity at the tip of the prosthesis on the static scan, and increased activity also seen associated with the right acetabulum. There is no increased vascularity at either of these sites.

Q3. Appearances suggest loosening of both the acetabular and femoral components of the hip prosthesis.

TEACHING POINTS

The most characteristic appearance of hip prosthesis loosening is increased uptake in the lesser trochanter and tip of the femoral component, usually without increased blood pool. There may be persistent acetabular and greater trochanter uptake for some years after insertion of a cementless prosthesis, which should be recognized as normal variants.

FURTHER READING

Rubello, D., Borsato, N., Chierichetti, F. *et al.* (1995) Three-phase bone scintigraphy patterns of loosening in uncemented hip prostheses. *Eur. J. Nuc. Med.*, **22**, 299–301.

Case 45

HISTORY

A six-year-old child presented with a history of urinary tract infections.

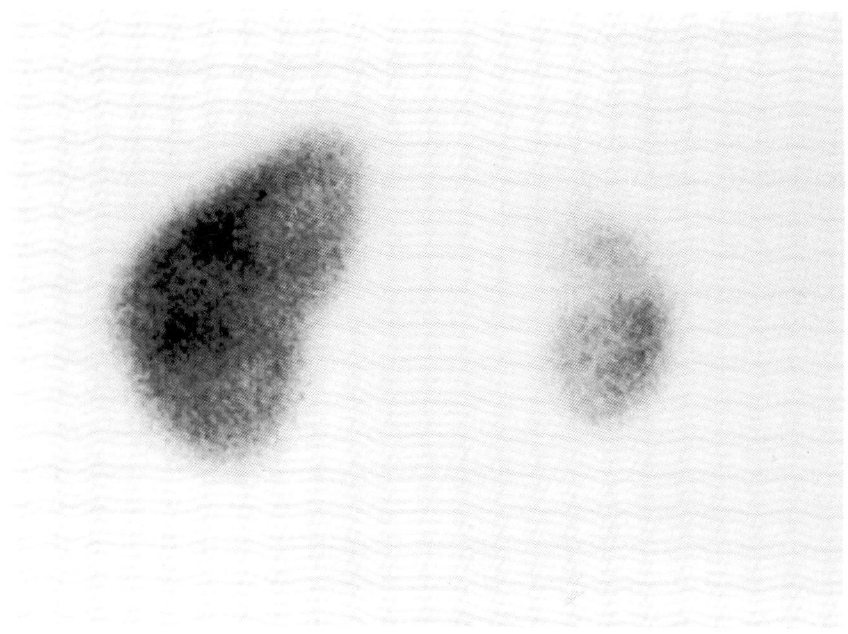

QUESTIONS

Q1. What is the examination?
Q2. What is the isotope used?
Q3. Describe the abnormalities and give a possible diagnosis.

ANSWERS

Q1. Static renal scan, posterior view.
Q2. ^{99m}Technetium dimercaptosuccinic acid (DMSA).
Q3. The right kidney is small compared with the left, and there is focal scarring of the upper pole. The right kidney contributes only 20% to the total glomerular filtration rate (GFR). The left kidney is essentially normal. Findings were due to previous reflux on the right with scarring and loss of function.

TEACHING POINT

Acute infection at the time of the scan needs to be excluded because a focal pyelonephritis can mimic a scar. This change, however, may be transient, and will usually resolve within about three months.

Case 46

HISTORY

A 56-year-old woman presented with metastatic right axillary lymph nodes. Her mammogram and breast ultrasound were normal. A MIBI study of the breasts was performed.

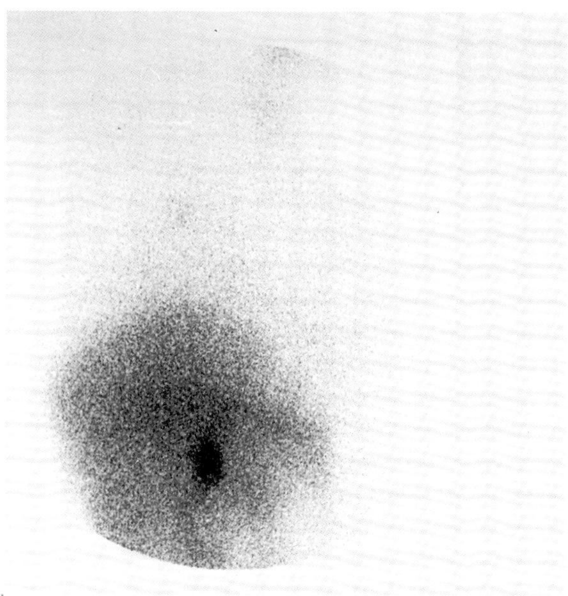

QUESTIONS

Q1. What does the scan show?
Q2. What does it mean?

ANSWERS

Q1. Focal uptake of the tail of the right breast.
Q2. Carinoma of the breast.

TEACHING POINT

Both thallium and MIBI have been shown to have a high sensitivity and specificity for the detection of carcinoma of the breast. Their roles in clinical practice are currently undefined, but they are likely to be an aid to diagnosis in patients with equivocal or difficult mammogram and/or biopsy results.

Case 47

HISTORY

A 40-year-old patient presented with a tender neck and underwent a technetium thyroid scan.

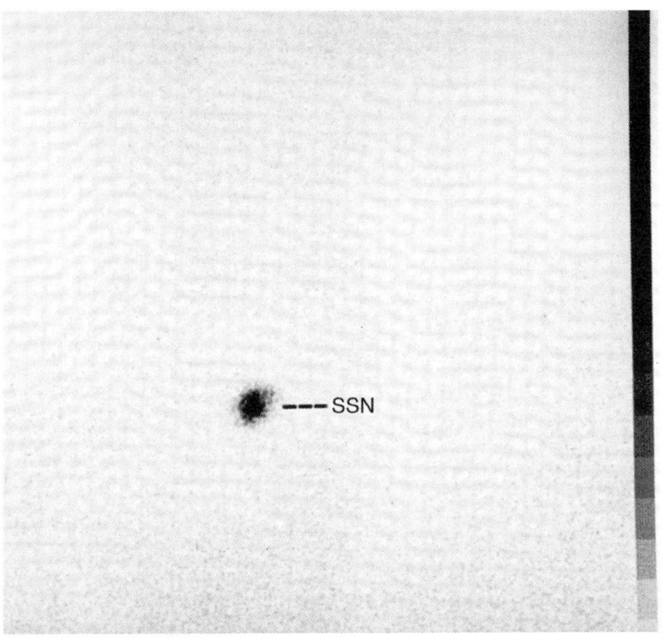

QUESTIONS

Q1. What does 'SSN' mean, and why is such a marker used?
Q2. What is the abnormality?
Q3. What is the probable diagnosis?

ANSWERS

Q1. 'SSN' stands for suprasternal notch marker. This is used to assist in demonstrating the relative position for the gland and any retrosternal extension.
Q2. There is no significant tracer uptake by the thyroid.
Q3. In view of the clinical history, the appearances are most likely to be due to subacute thyroiditis.

TEACHING POINTS

1. Although less likely, other diagnostic possibilities include thyroxine use and iodine loading, e.g. with radiological contrast or drugs such as amiodarone, which has a 40% iodine content.
2. Less likely causes include:

   - ectopic thyroid tissue
   - high intake of iodine in food substances
   - surgery.

3. A repeat thyroid scan in two to three months will generally show resolution in subacute thyroiditis.

Case 48

HISTORY

A 36-year-old woman presented with jaundice. The cause was unknown. A HIDA (dimethyl iminodiacetic acid) study was performed.

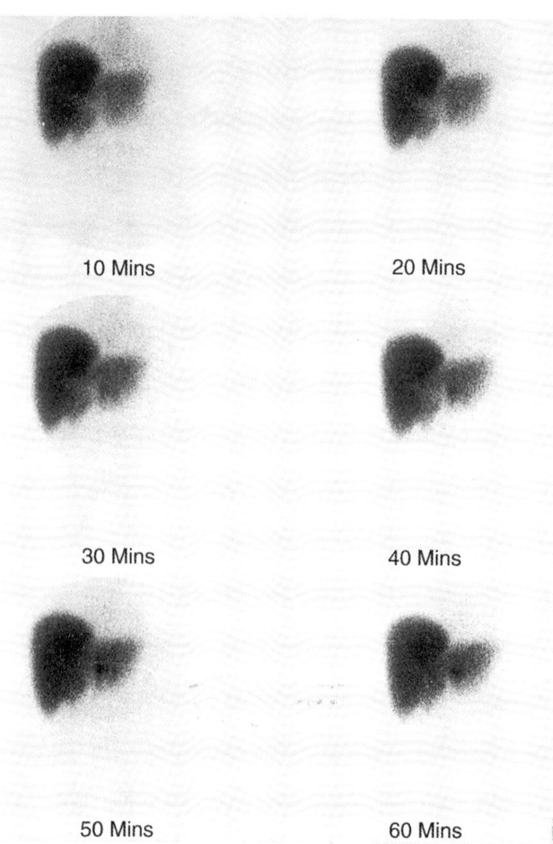

QUESTIONS

Q1. What does the scan show?
Q2. What does this imply?

ANSWERS

Q1. Uptake in liver with no biliary excretion (colloid appearance).
Q2. Obstructed jaundice.

Case 49

HISTORY

A 71-year-old man presented with weight loss and bone pain. A whole body bone scan was performed.

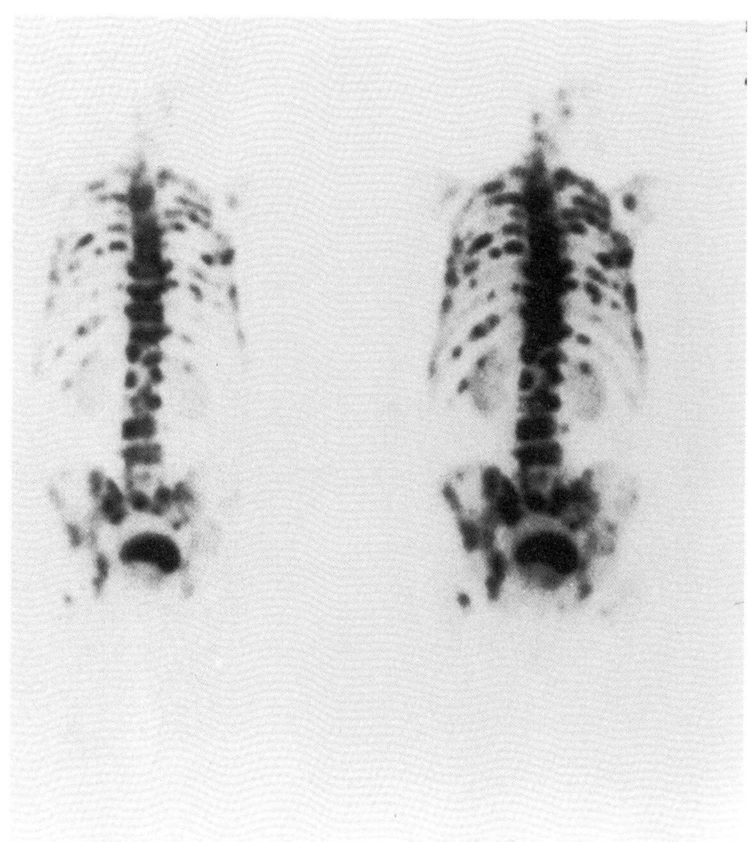

QUESTIONS

Q1. Describe the abnormalities.
Q2. Give your interpretation of the findings.

ANSWERS

Q1. Multiple focal areas of abnormal uptake are present throughout the skeleton. Abnormal sites include the ribs bilaterally, the thoracic and lumbar vertebrae, the sacrum and pelvis. Further lesions are also seen in the left scapula, the neck, and the right femur.

Q2. Scan findings are those of multiple metastases. There is a filling defect at the base of the bladder, which suggests carcinoma of the prostate as the underlying primary malignancy.

Case 50

HISTORY

A 55-year-old female presented with aches and pains. A whole body scan was performed.

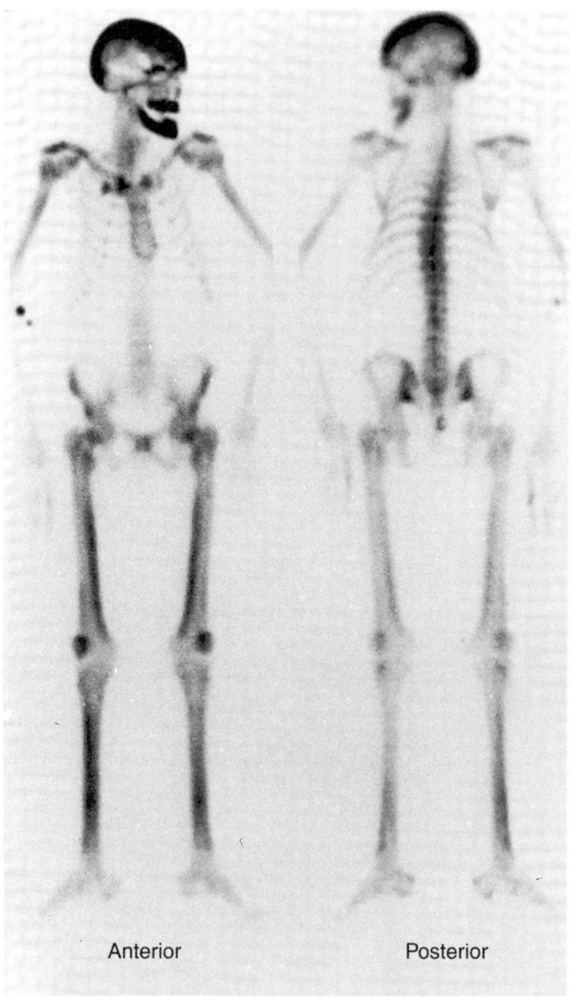

QUESTIONS

Q1. Describe the scan.
Q2. What is your interpretation?

ANSWERS

Q1. There is striking increased uptake of tracer throughout the whole of the skeleton, with heightened contrast between the bone and soft tissues. Note the absence of renal images. There are additional metabolic features with increased uptake throughout the calvaria, the mandible, the costochondral junctions bilaterally, and the sternum (tie sign).

Q2. The appearances are those of a super scan due to metabolic bone disease. Renal osteodystrophy is the most likely cause.

TEACHING POINT

A super scan of malignancy generally shows some more focal areas of uptake. The skull appearances in this case are virtually pathognomonic of metabolic bone disease, and are only seen in the presence of severe hyperparathyroidism.

Case 51

HISTORY

A 40-year-old man presented with hip pain and nephrotic syndrome.

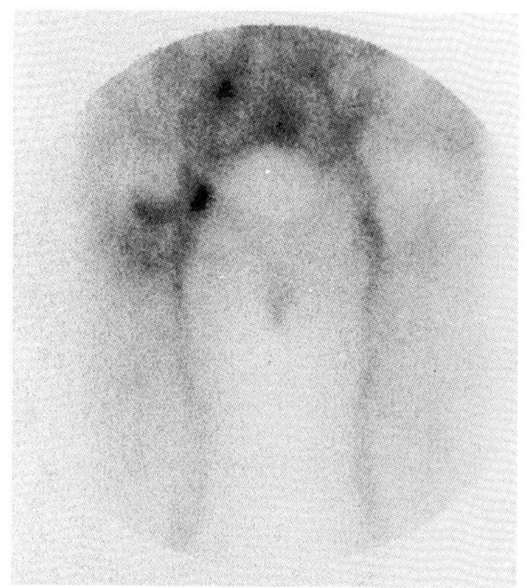

Blood pool.

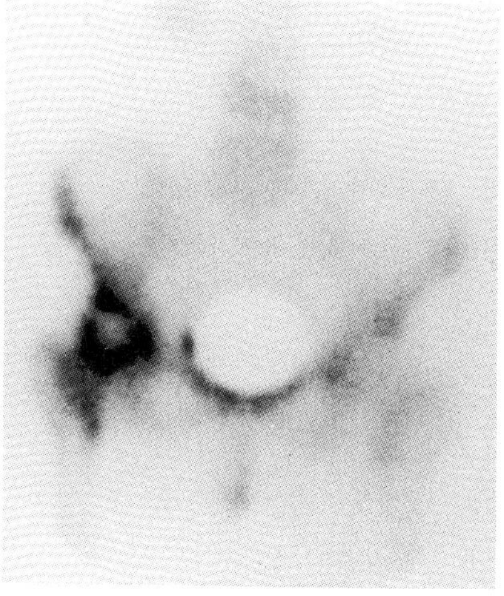

Static delayed image.

QUESTIONS

Q1. Describe the abnormalities.
Q2. What is the most likely diagnosis?
Q3. Give a differential diagnosis of the possible underlying aetiology.

ANSWERS

Q1. There is increased uptake of tracer in the right hip on both the blood pool and static images. In addition, a striking central photon deficient area associated with the right hip can be seen.

Q2. The appearances are most likely to be those of avascular necrosis of the right hip relating to the patient's steroid therapy for nephrotic syndrome.

Q3. Other causes of avascular necrosis include:

- cytotoxic therapy
- radiotherapy
- alcoholism
- trauma
- sickle cell disease
- pancreatitis
- Gaucher's disease and similar haematological disorders
- idiopathic.

TEACHING POINT

In this context, the central photon deficient lesion assists in the diagnosis, but it is sometimes not seen on planar images. SPECT should be performed routinely in this situation as it can enable detection of a central photon deficient area and probable avascular femoral head.

Case 52

HISTORY

A 20-year-old jogger presented with bilateral leg pain.

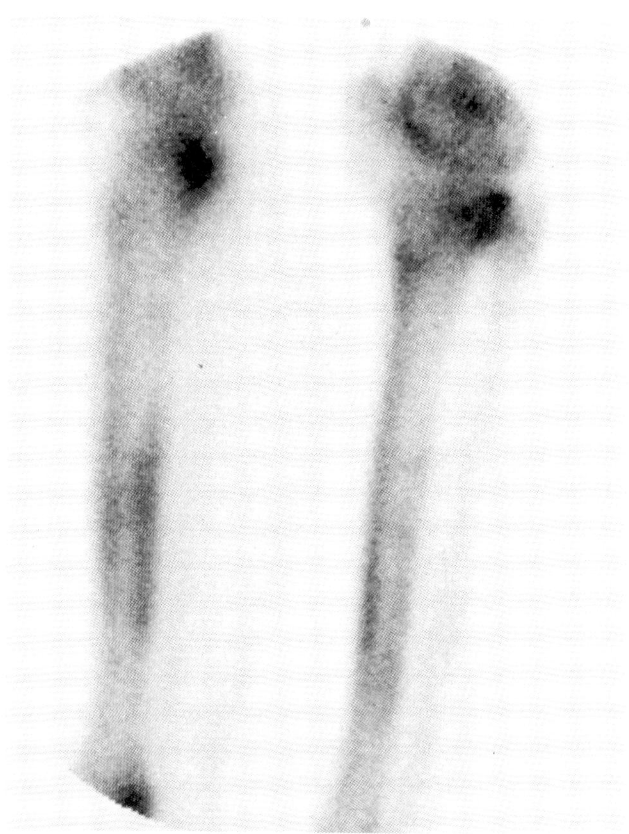

QUESTIONS

Q1. What is the abnormality?
Q2. What is the likely diagnosis?
Q3. What is the underlying pathology?
Q4. Can this patient continue jogging?

ANSWERS

Q1. There is increased uptake diffusely along the posterior lower third of the right tibia with some increased tracer uptake throughout the rest of the tibiae.

Q2. 'Shin splints'.

Q3. This is a syndrome that is due to a periosteal reaction at the site of muscle insertions at the lower third of the posterior tibia.

Q4. The patient can continue jogging if he feels comfortable. This is in contrast to a patient with a stress fracture, who must rest in order to prevent progression to a complete fracture.

TEACHING POINTS

There is no increased vascularity associated with shin splints, unlike the appearances seen with a stress fracture. Lateral views of the lower limbs are needed in patients presenting with leg pain, otherwise the diagnosis of shin splints will be missed.

Case 53

HISTORY

A 27-year-old male patient presented with back pain. Ankylosing spondylitis was suspected. Posterior views of the pelvis on the patient's bone scan are shown.

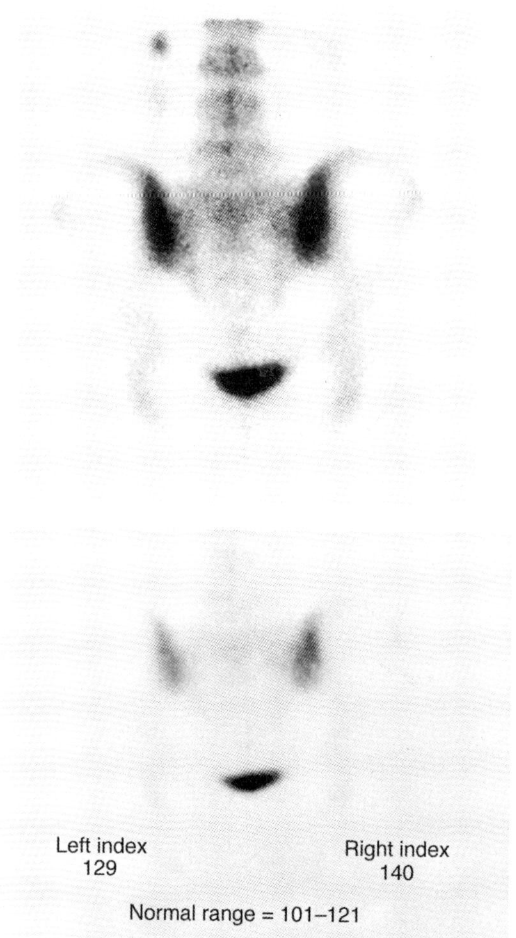

QUESTIONS

Q1. Does this show any abnormality?

Q2. How can additional processing of the completed image assist in reaching a diagnosis?

ANSWERS

Q1. There is increased uptake in the sacroiliac joints. The appearances are consistent with sacroiliitis. With the patient's clinical history, ankylosing spondylitis is the most likely underlying cause.

Q2. The regions of interest selected over the sacroiliac joint and sacrum will give an index reflecting the ratio for counts between the sacroiliac joints and sacrum, as seen with this sacroiliac quantitation. Both sacroiliac joints show increased uptake, more on the right.

TEACHING POINTS

1. A bone scan is more sensitive than plain films in the early diagnosis of sacroiliac disease because it reflects altered metabolic activity.
2. Typical causes of a bilateral symmetrical sacroiliitis include ankylosing spondylitis, inflammatory bowel disease, and psoriatic arthropathy.
3. Rarer causes of sacroiliitis include:

   - Reiter's syndrome
   - rheumatoid arthritis
   - gout.

Case 54

HISTORY

A female patient being treated for carcinoma of the breast had a bone scan.

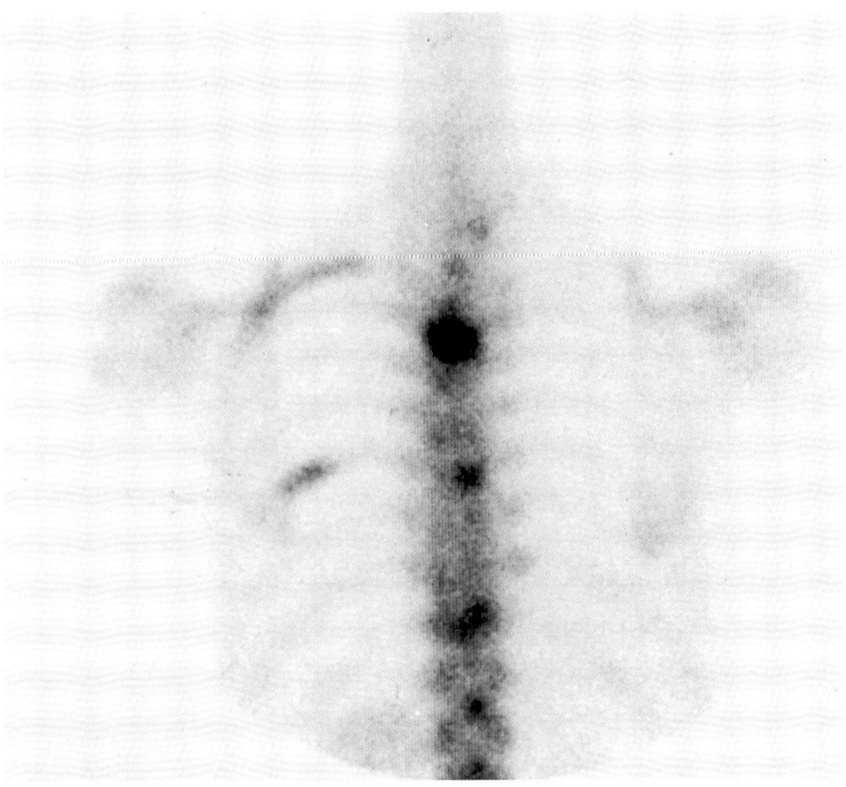

QUESTIONS

Q1. Describe the abnormalities.
Q2. Give a possible explanation.

ANSWERS

Q1. On the initial scan there are areas of increased uptake related to the right 3rd and 7th ribs posteriorly, but also the bodies of T4, T7 and T10.

Q2. The appearances are those of metastases.

TEACHING POINT

Uptake extending along a rib is highly suggestive of malignancy. Response to treatment with partial resolution of abnormalities is uncommon, but can be seen in this case (see below).

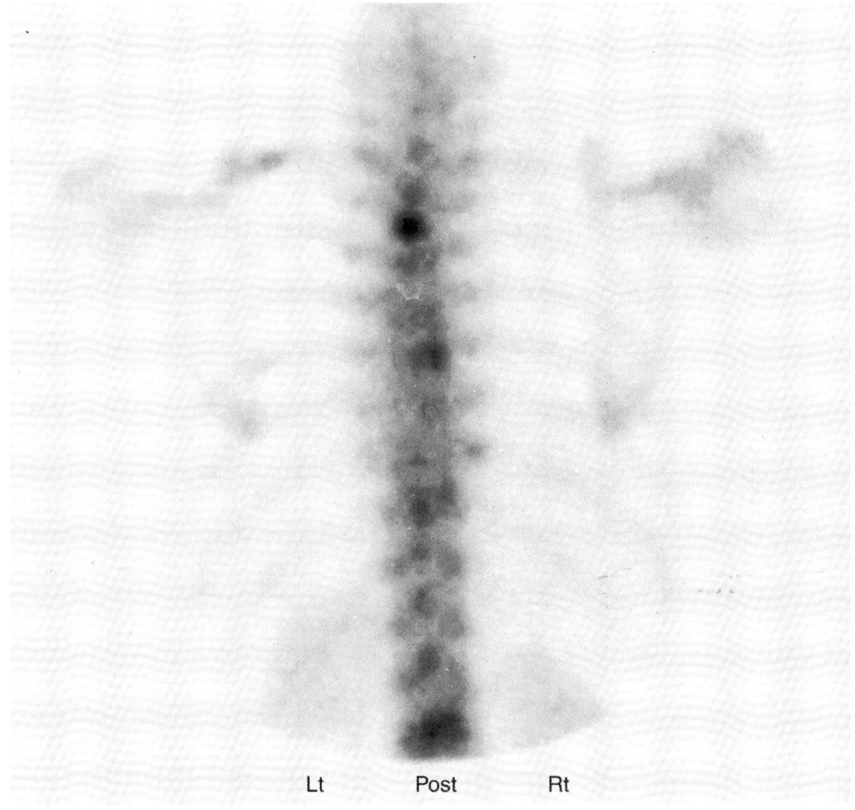

Case 55

HISTORY

A 40-year-old man presented with posterior chest pain (the remainder of the bone scan is normal).

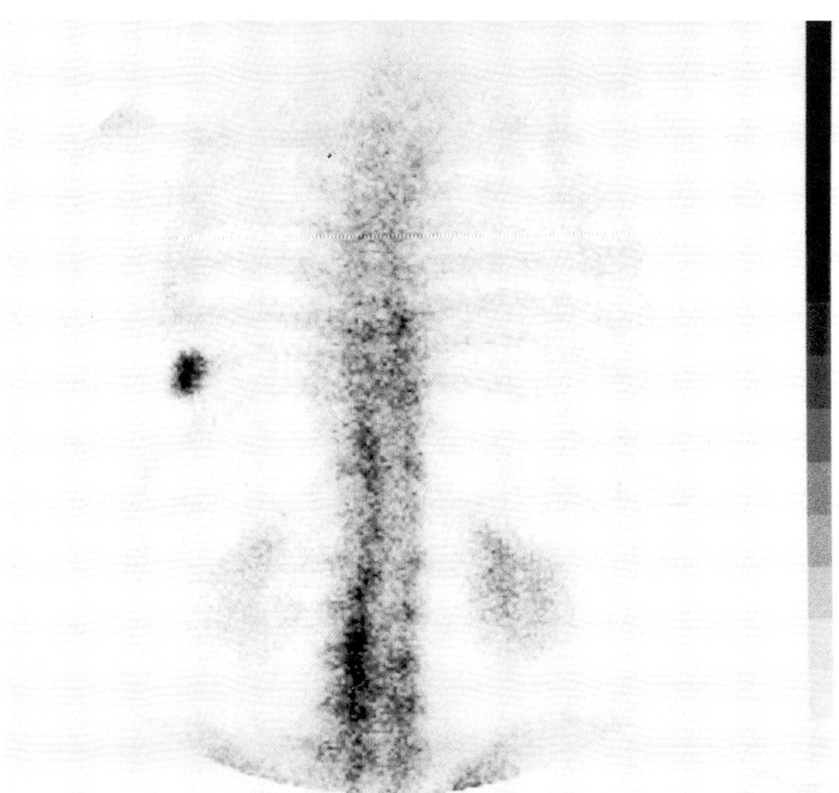

QUESTIONS

Q1. What is the abnormality?
Q2. What would be your next investigation?
Q3. Give a differential diagnosis.

ANSWERS

Q1. There is a solitary focal area of increased uptake in the left 8th rib posteriorly.

Q2. An x-ray of this area should be performed to exclude bony destruction at this site. The clinical history should also be taken to exclude a history of trauma.

Q3. The differential diagnosis of a solitary rib lesion includes:

- a metastasis (a solitary rib lesion has a 10% probability of being a metastasis, although this rises to 50% in a patient with known pre-existing malignancy)
- trauma
- a benign tumour.

Case 56

HISTORY

A patient with bilateral hip replacements (the right having been performed five years previously and the left seven) presented with pain in the left hip.

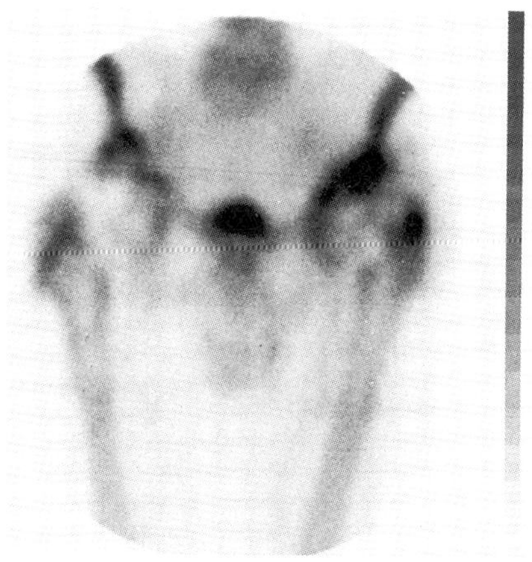

Delayed anterior bone scan.

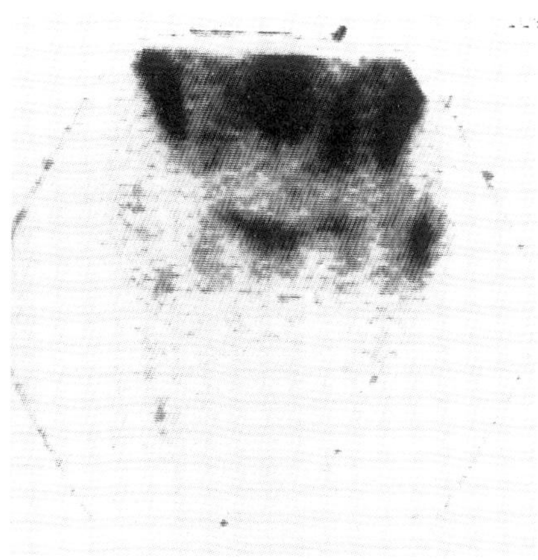

Indium white cell scan at 24 hours post-injection.

QUESTIONS

Q1. Describe the abnormalities.
Q2. Give a likely diagnosis.

ANSWERS

Q1. Bilateral total hip replacements are demonstrated with increased bone scan uptake. An area of markedly increased uptake on the indium white cell scan is seen to be associated with the left hip prosthesis in the region of the greater trochanter.

Q2. The appearances are most likely to represent infection related to the left total hip replacement.

TEACHING POINT

To be considered positive for infection, indium white cell imaging should show uptake of tracer that is similar to, or more extensive than, the corresponding bone scan lesion.

Case 57

HISTORY

A bone scan was performed in an osteoporotic patient who complained of low back/buttock pain. The posterior view is demonstrated.

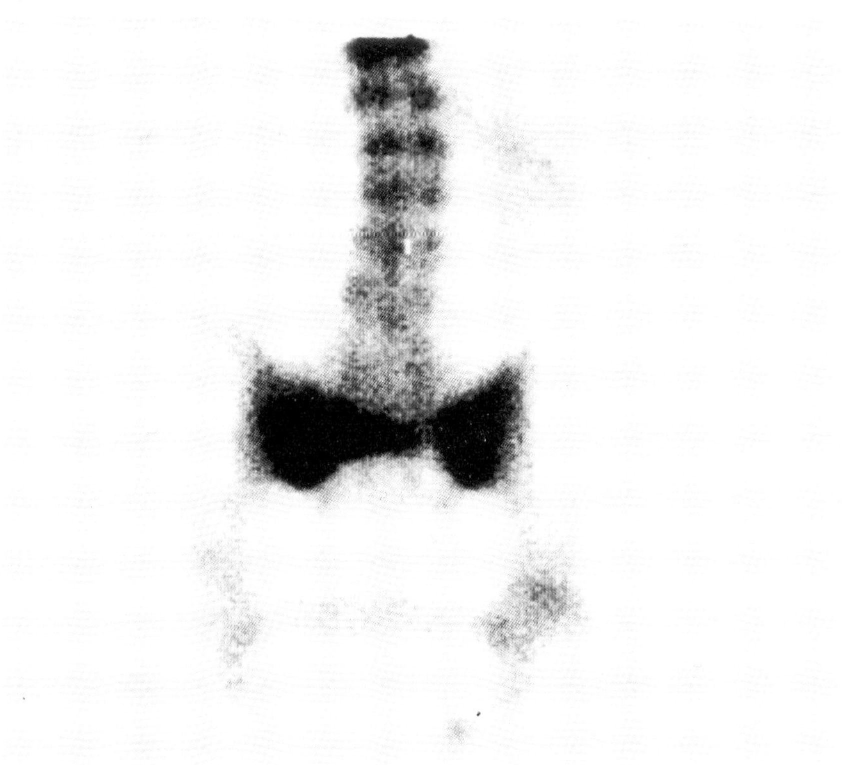

QUESTIONS

Q1. Describe the abnormalities.
Q2. What is the probable diagnosis?

ANSWERS

Q1. There is increased uptake throughout S1 and the sacroiliac joints on both sides, with further increased uptake seen in the lower thoracic vertebrae and a small focus in the right inferior pubic ramus.

Q2. Spinal and pelvic fractures. The clinical history revealed a recent fall and fractures were confirmed on x-ray.

TEACHING POINT

The classic appearance of a pelvic insufficiency fracture in osteoporosis is an 'H' appearance, with the vertical limbs of the 'H' representing the sacroiliac joints and the horizontal part the sacrum. Incomplete 'H' appearances can also be seen.

Case 58

HISTORY

A 52-year-old man presented with fits and ataxia. A CT scan showed ventricular dilatation. A radionuclide study was performed. Images are displayed at six hours, 24 hours and 48 hours after injection.

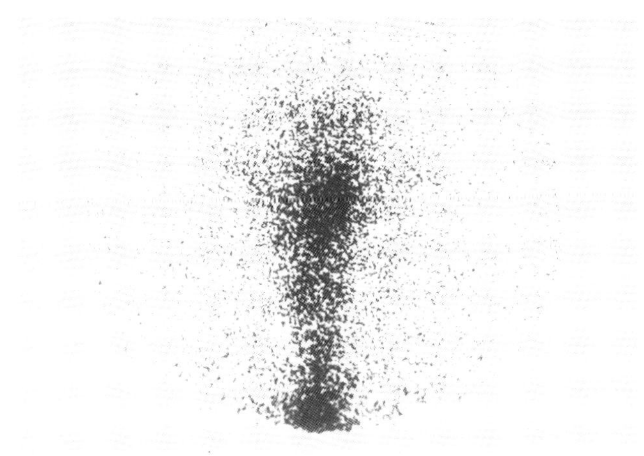

Six-hour anterior view.

24-hour anterior view.

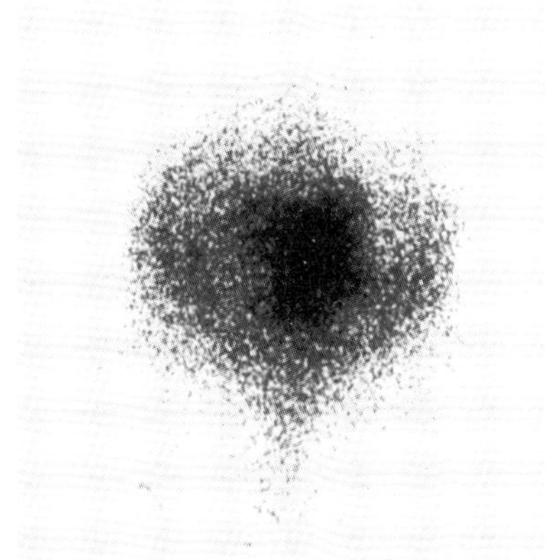

48-hour anterior view.

QUESTIONS

Q1. What is the study?
Q2. What is the abnormality?

ANSWERS

Q1. ^{111}In DTPA cisternogram.

Q2. There is reflux of tracer into the ventricles with retention in the ventricles at 48 hours. This is typical of communicating hydrocephalus.

TEACHING POINT

The diagnosis of communicating hydrocephalus requires the demonstration of reflux of tracer into the ventricles and retention within them. Emptying of the ventricles would normally be expected by 48 hours.

FURTHER READING

Gilday, D.L. (1991) Special clinical problems in paediatrics, in *Clinical Nuclear Medicine* (eds M.N. Maisey, K.E. Britton and D.L. Gilday), Chapman & Hall, London, pp. 413–18.

Case 59

HISTORY

A 35-year-old woman presented with renal calculi. Hypercalcaemia was found on investigation and the serum parathormone was also raised. A parathyroid localization study was performed.

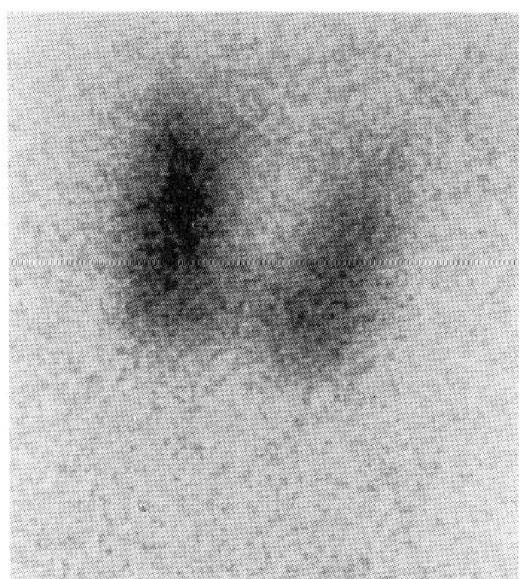

Pertechnetate scan.

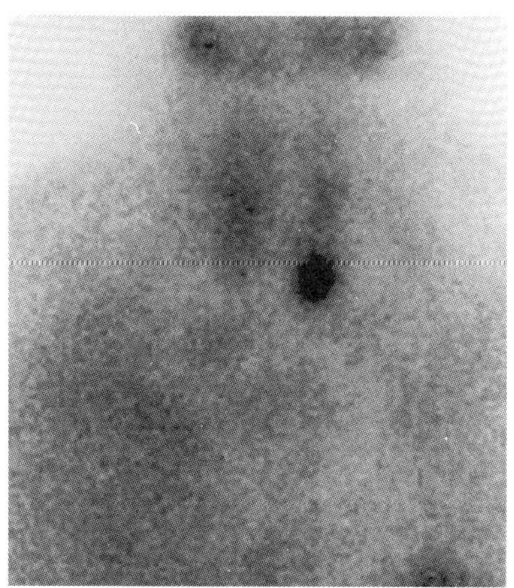

Delayed MIBI study.

QUESTIONS

Q1. What does the study show?

Q2. How does a MIBI localization study differ from a technetium–thallium subtraction scan?

ANSWERS

Q1. A left sided parathyroid adenoma.

Q2. Technetium–thallium subtraction imaging has been established as the optimum non-invasive localization study for a parathyroid adenoma. MIBI is a widely used alternative to thallium for myocardial perfusion imaging and has now been shown to have applications for tumour imaging.

In parathyroid adenomas, there is slower washout of the sestamibi from the adenoma compared with the remainder of the thyroid and imaging is performed 20–30 minutes and three hours after injection.

FURTHER READING

O'Doherty, M.J., Kettle, A.G., Wells, P. *et al*. (1992) Parathyroid imaging with technetium 99m sestamibi: preoperative localisation and tissue uptake studies. *J. Nuc. Med.*, **33**, 313–18.

Taillefer, R., Boucher, Y., Patrin, C. and Lambert, R. (1992) Detection and localisation of parathyroid adenomas in patients with hyperparathyroidism using a single radionuclide imaging procedure with technetium 99m sestamibi (double phase study). *J. Nuc. Med.*, **33**, 1801–9.

Case 60

HISTORY

A 63-year-old woman developed poorly controlled hypertension. Clinical examination was normal; serum biochemistry including electrolytes and calcium was normal; VMAs were normal. A captopril MAG 3 (technetium-99 mercapto-acetyltriglycine) study was performed.

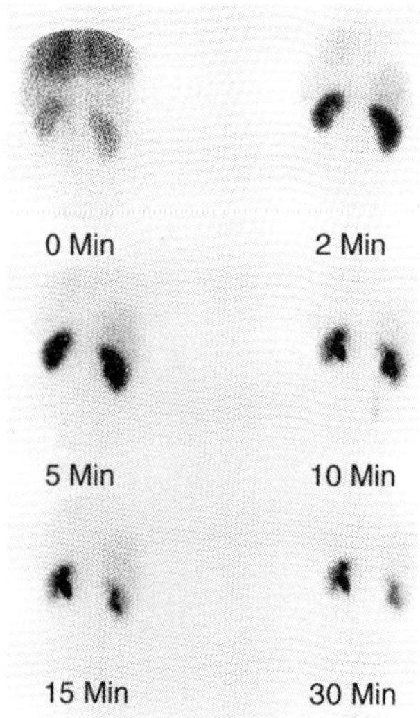

MAG 3 Study without captopril. Images at zero, two, five, 10, 15 and 30 minutes post-injection.

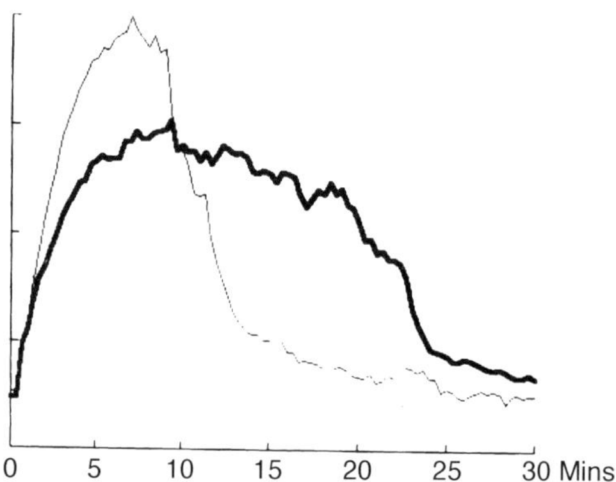

Renogram curve (left kidney, dark line). The left kidney is estimated to contribute 46% to the divided function from background subtracted counting between one and two minutes post-injection.

QUESTIONS

Q1. What does the study demonstrate?
Q2. How does captopril work in this test?
Q3. What further investigation should be performed?

ANSWERS

Q1. There is a decline in function of the left kidney following the administration of captopril, suggesting a diagnosis of renal artery stenosis (RAS).

Q2. In RAS, perfusion of the glomeruli is maintained by increased efferent artery tone under angiotensin II control. Blockage of the synthesis of angiotensin II with captopril lowers the efferent artery tone; glomerulus perfusion consequently declines and renal function worsens.

Q3. A positive captopril MAG 3 study means significant renovascular hypertension is present, although this can be small or large vessel disease. Digital subtraction angiography or arteriography is the next appropriate test. A positive captopril MAG 3 study is strongly associated with a good outcome to arterial dilatation provided that the contralateral kidney is normal.

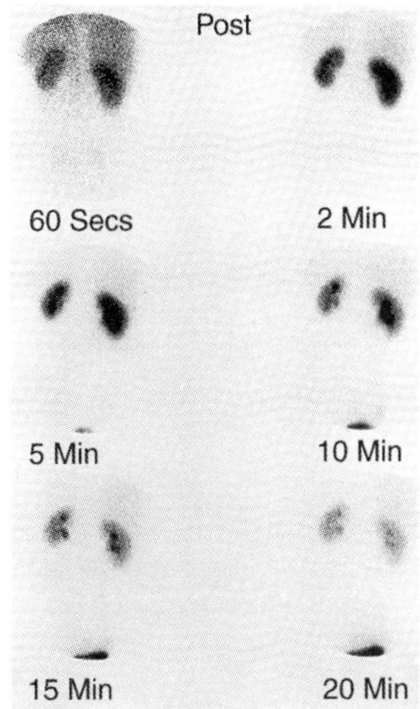

MAG 3 study after captopril administration. Images obtained at one, two, five, 10, 15 and 20 minutes post-injection.

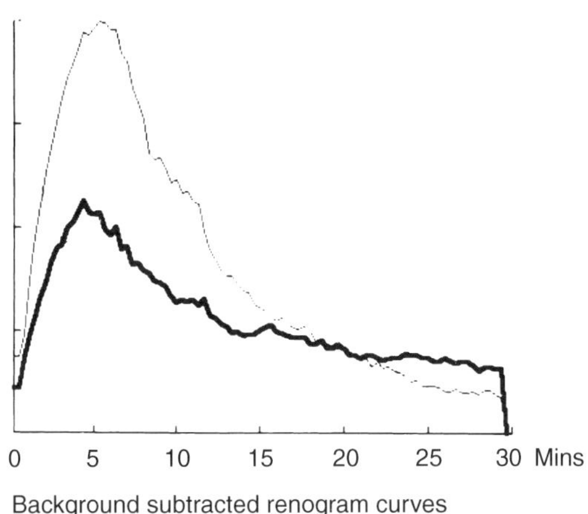

Renogram curve. Divided function: left, 37%; right, 63%.

FURTHER READING

Tulchin, M., Eggli, F., Waybill, P. and Diamond, J.R. (1993) Diagnosis of neuro-vascular hypertension: nuclear medicine techniques. *App. Radiol.*, **22**, 13–20.

Case 61

HISTORY

A four-year-old girl had persistent urinary tract infection. An indirect reflux study was performed to identify whether reflux was present. Images from the reflux study with background subtracted count profiles over bladder and kidneys are displayed.

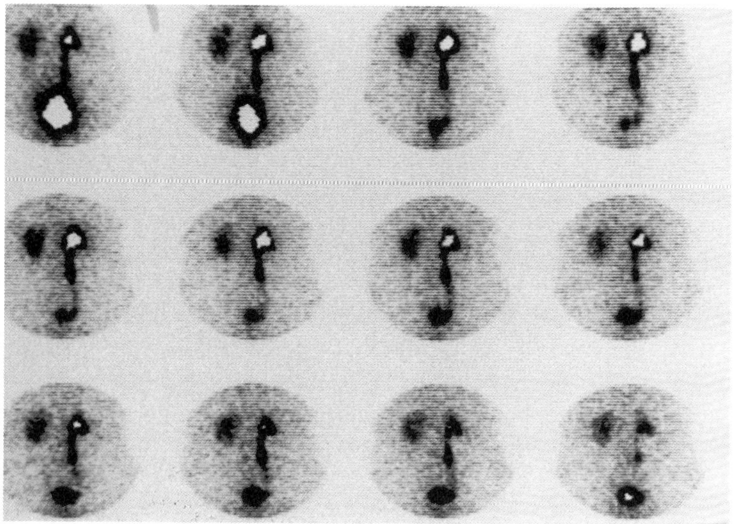

Images of the kidneys and bladder during the reflux study.

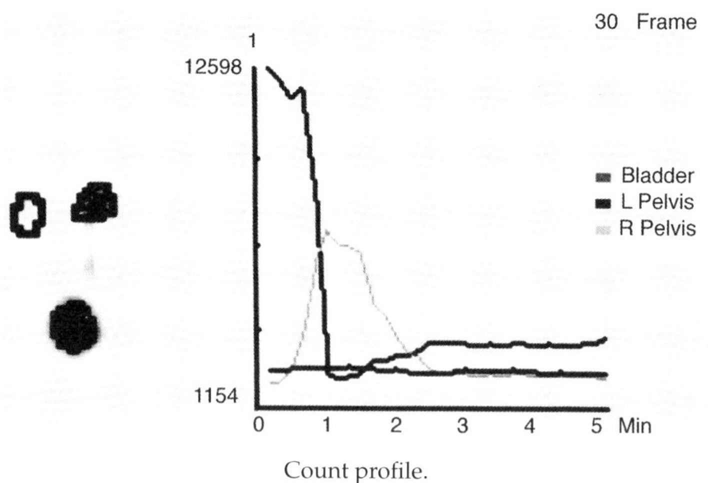

Count profile.

QUESTIONS

Q1. Is the study satisfactory?
Q2. What is the abnormality?

ANSWERS

Q1. Yes. There is excellent bladder emptying as judged by a rapid fall in the bladder curve. Both kidneys and bladder are clearly visualized.

Q2. There is a rise in activity over the right kidney as the bladder empties, implying reflux to that kidney. This can be also seen on the images where there is increased activity in the right kidney, which then empties.

TEACHING POINT

Dynamic renal imaging with indirect reflux offers a sensitive, low radiation dose physiological approach to the detection of vesicoureteric reflux but is restricted to individuals with bladder control. For children without bladder control, a direct radionuclide or contrast reflux study is required.

FURTHER READING

Pollet, J.E., Sharp, F. and Smith, R.W. (1979) Radionuclide imaging for detecting renal reflux using intravenous 99m Tc DTPA. *Paediat. Radiol.*, **8**, 165–7.

Case 62

HISTORY

An 82-year-old woman presented with confusion. A brain scan was performed to exclude serious pathology.

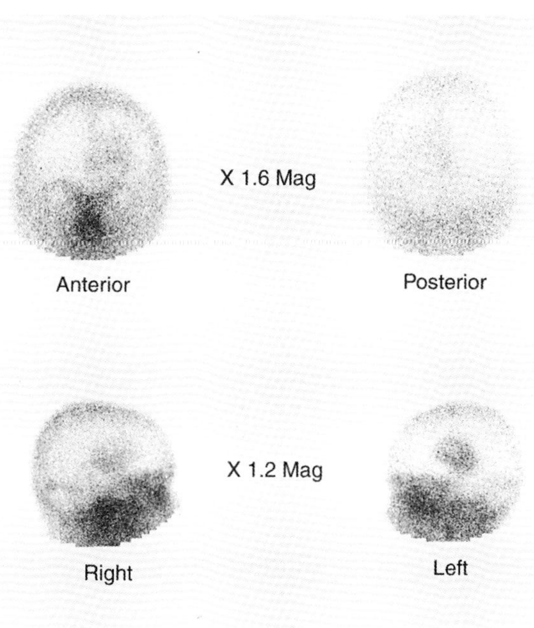

QUESTIONS

Q1. What type of brain scan is this?
Q2. What does it show?
Q3. What are the possible causes?

ANSWERS

Q1. A planar brain scan using an agent such as DTPA or glucoheptonate that only crosses the blood–brain barrier where pathology is present.
Q2. A right-sided central brain lesion close to the mid-line.
Q3. Tumour, most probably a meningioma.

TEACHING POINT

Blood–brain barrier agents are very sensitive for the detection of pathology such as haemorrhage, stroke or tumour. This makes such a simple test a good technique for screening confused elderly patients. The test is not now commonly performed as the lack of specificity means that a CT scan is preferable in most situations.

Case 63

HISTORY

A 45-year-old man had previous parathyroid surgery for primary hyperparathyroidism. No adenoma was found but three and a half glands were removed at surgery. The patient remained hypercalcaemic and a combined pertechnetate and MIBI study was performed.

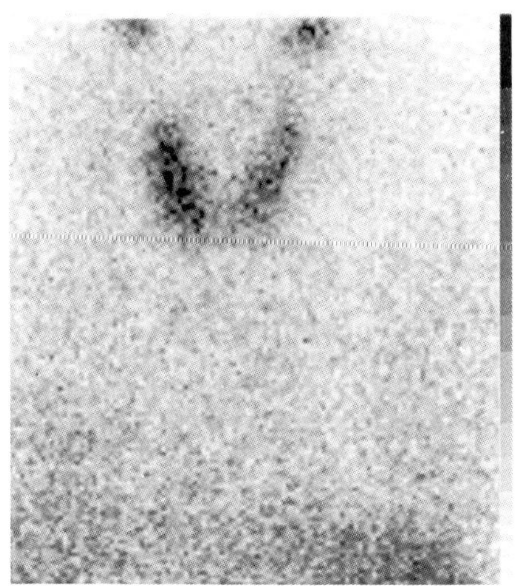

Pertechnetate thyroid scan.

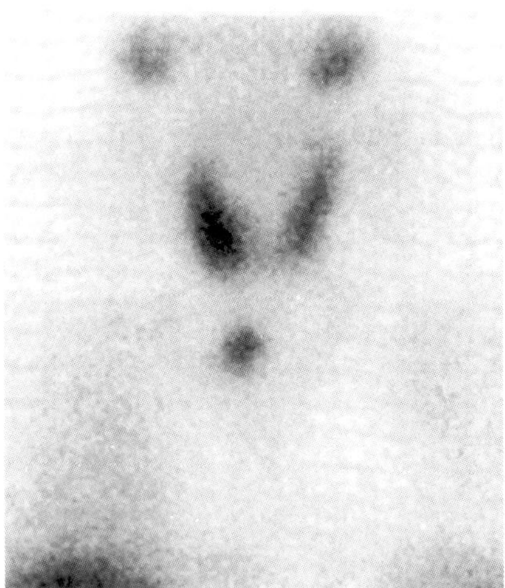

MIBI thyroid scan.

QUESTIONS

Q1. What is the abnormality?
Q2. What is the cause?

ANSWERS

Q1. The pertechnetate thyroid scan is normal, but the MIBI study shows a focus of increased activity below the thyroid in the upper mediastinum.

Q2. Mediastinal parathyroid adenoma.

TEACHING POINT

Nuclear medicine imaging will detect approximately 95% of parathyroid adenomas. However, negative studies may occur with small adenomas, hyperplasia (10–15% of cases of primary hyperparathyroidism) and adenomas in ectopic sites. It is important to include the mediastinum in the field of view when performing parathyroid imaging.

Case 64

HISTORY

A 70-year-old man presented with lower GI bleeding. Upper GI endoscopy was normal and the patient continued to bleed. A red blood cell (RBC) 'bleeding' study was performed. Anterior images are shown at 0 minutes, three hours, five hours, and five hours 15 minutes after injection.

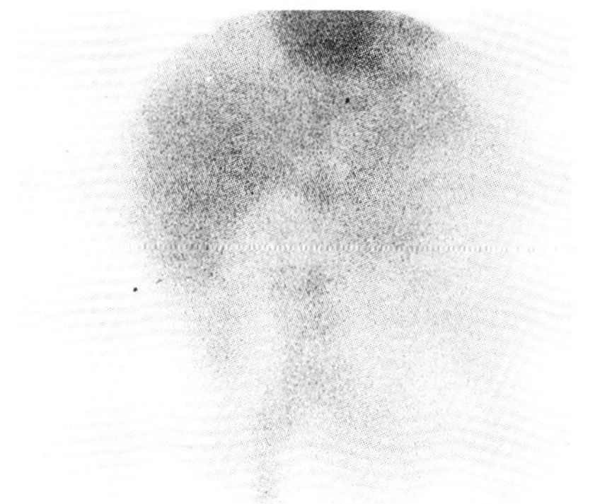

Zero minutes.

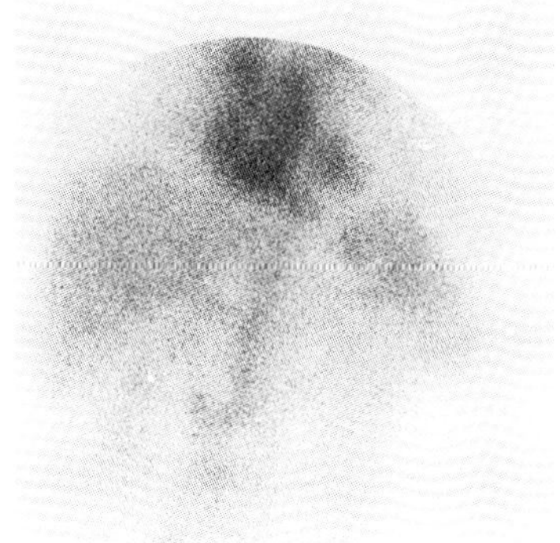

Three hours post-injection.

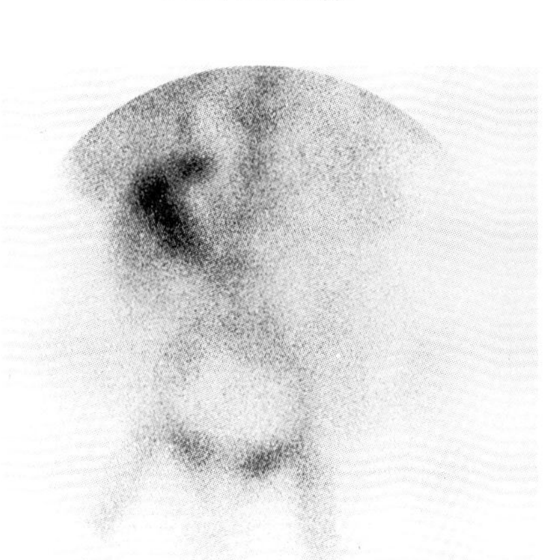

Five hours post-injection.

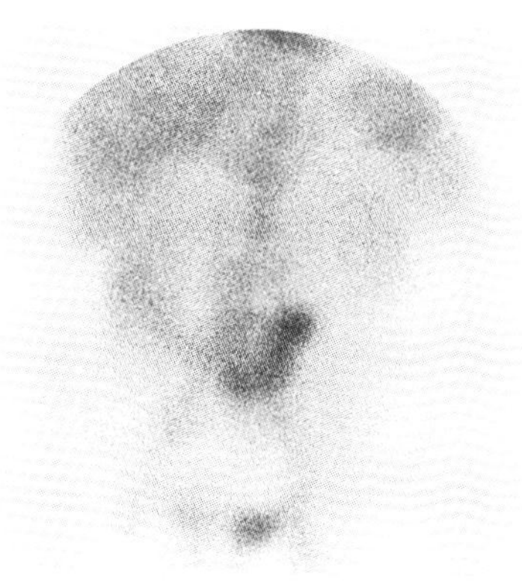

Five-and-a-quarter hours post-injection.

QUESTIONS

Q1. What abnormality is shown?
Q2. How should it be interpreted?

ANSWERS

Q1. Images taken immediately post-injection show a normal blood pool. The study remains normal at three hours. However, by five hours, abnormal activity is visualized in the right upper quadrant, and by five and a quarter hours this can be seen to have moved inferomedially.

Q2. The study is positive for active bleeding, with the bleeding site most probably being in small bowel.

TEACHING POINT

A positive study is likely to be obtained with a bleeding rate greater than 0.5 ml/minute. As in the present study, it is often worth continuing imaging for several hours post-injection to identify a bleeding source. It must be remembered that if bleeding is profuse, the labelled RBCs may have moved some distance from the site of bleeding by the time of imaging unless imaging is continuous.

FURTHER READING

Winzelberg, G.S., McKusick, K.A., Strauss, A.L. *et al.* (1979) Evaluation of gastrointestinal bleeding by red blood cells *in vivo* with technetium 99m. *J. Nuc. Med.*, **20**, 1080–6.

Case 65

HISTORY

A 64-year-old man with hypercalcaemia was investigated with a whole body bone scan. A posterior lumbar spine image is displayed.

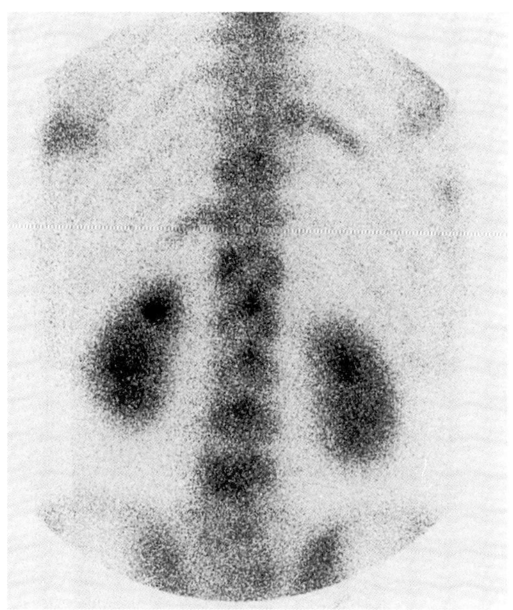

QUESTIONS

Q1. Describe the abnormalities.
Q2. What do they imply?

ANSWERS

Q1. Multiple focal increased activity in the lower thoracic and lumbar spine. There are also multiple rib foci with activity extending along the left 1st and right 9th ribs. Increased activity is also present in the left scapula and both kidneys.

Q2. The bone lesions are typical for skeletal metastases. Renal activity is frequently increased in the bone scan in the presence of hypercalcaemia.

TEACHING POINT

High renal retention of bone disphosphonate tracers is most commonly seen in hypercalcaemia or in patients being treated with cytotoxic drugs.

Case 66

HISTORY

A whole body scan was performed on a 59-year-old man with low back pain. A posterior image of the spine and pelvis and an anterior view of the pelvis and femora is shown.

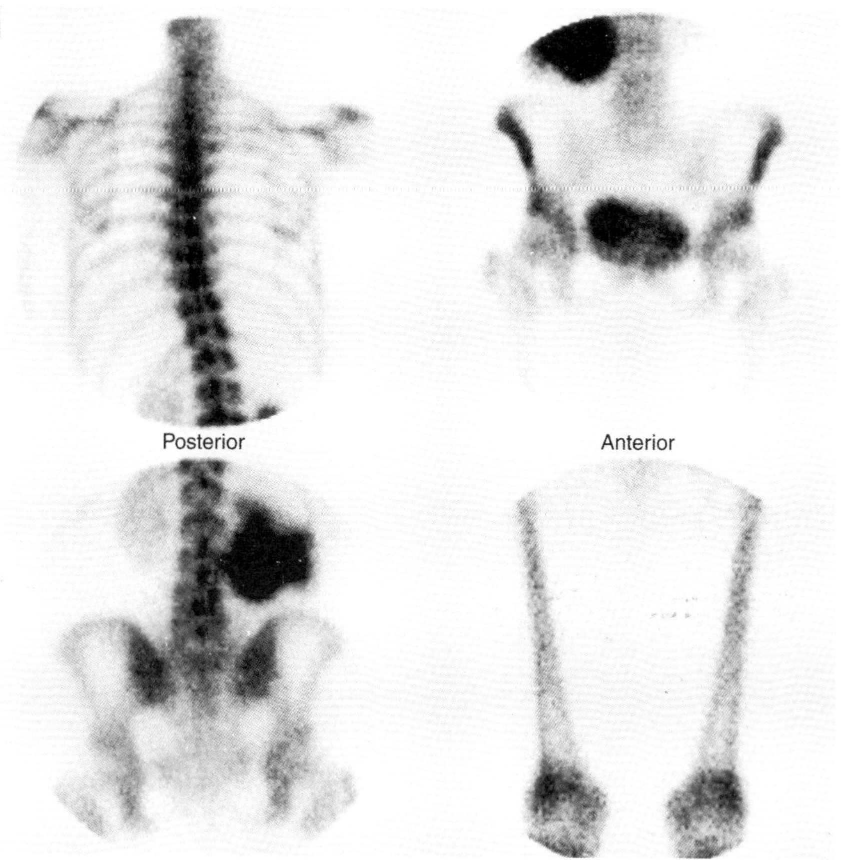

QUESTIONS

Q1. What is the abnormality?
Q2. What further investigations should be performed?

ANSWERS

Q1. There is massive accumulation of the bone agent in a dilated renal right pelvis. The remainder of the study is normal.

Q2. The study suggests a right PUJ obstruction. Further investigation with a DTPA study to confirm PUJ obstruction and an ultrasound scan to detect a structural abnormality would be worthwhile.

TEACHING POINT

Renal abnormalities are commonly found on bone scan. It is important to examine the kidneys routinely as well as the skeleton, as important diagnostic clues might otherwise be missed. The changes are dramatic in the present case, but more subtle abnormalities can occur.

Case 67

HISTORY

A patient developed anuria two days after insertion of a left-sided renal transplant. Images are shown at 30 seconds, one minute, five and 10 minutes after injection of ^{99m}Tc DTPA.

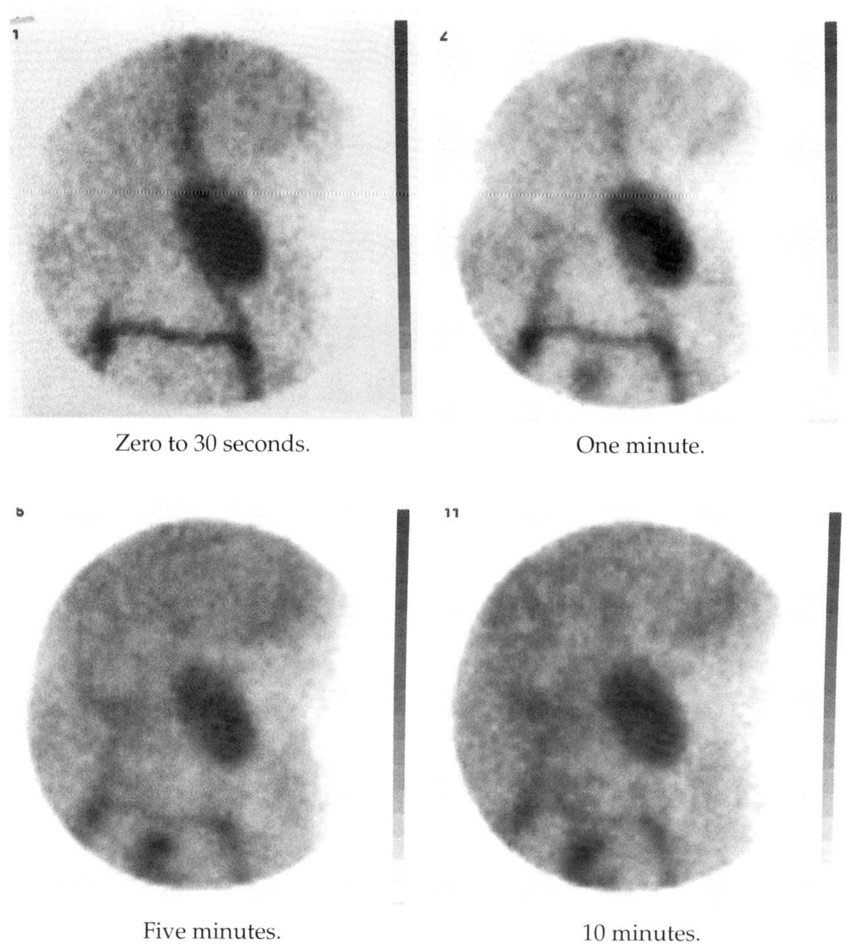

Zero to 30 seconds. One minute.

Five minutes. 10 minutes.

QUESTIONS

Q1. Describe the findings in this study.
Q2. What does this imply?

ANSWERS

Q1. There is very good blood flow to the left kidney and a good blood pool image at one minute. Subsequently there is no excretion and the blood pool fades.
Q2. These are typical changes of acute tubular necrosis.

TEACHING POINT

Scan findings in renal failure

Pre-renal	*ATN**	*Parenchymal*
Normal blood flow	Normal blood flow	Impaired blood flow
Good uptake	Absent or poor uptake	Poor uptake
Delayed intrarenal transit		Delayed transit
Minimal excretion	No excretion	Poor or absent excretion

* ATN, acute tubular necrosis.

Note that DTPA alone cannot be used to exclude obstruction in renal failure.

Case 68

HISTORY

A 36-year-old woman with rheumatoid arthritis presented with a fever and a painful left hip. A two-phase bone scan was performed.

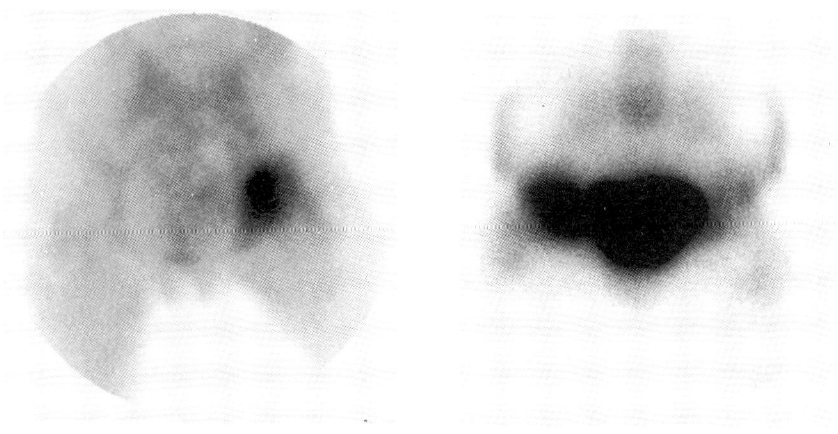

Blood pool (posterior). Static image (anterior).

QUESTIONS

Q1. What abnormalities are shown?
Q2. What is the most likely diagnosis?
Q3. What further tests would be appropriate?

ANSWERS

Q1. Florid increased activity in the left hip joint with markedly increased blood pool.

Q2. Septic arthritis. The activity is more intense than one would expect from inflammatory arthritis alone. Avascular necrosis should be considered, particularly if the patient is on steroids, but activity would then be expected to be in the femoral head rather than affecting both sides of the joint.

Q3. If a significant effusion is present, this could be detected by ultrasound and an ultrasound-guided aspiration performed. If there is doubt, a labelled white cell study should be performed.

Case 69

HISTORY

A 32-year-old woman had a polyarthritis. The diagnosis was unclear. A Tc human polyclonal immunoglobulin (HIg) study was performed.

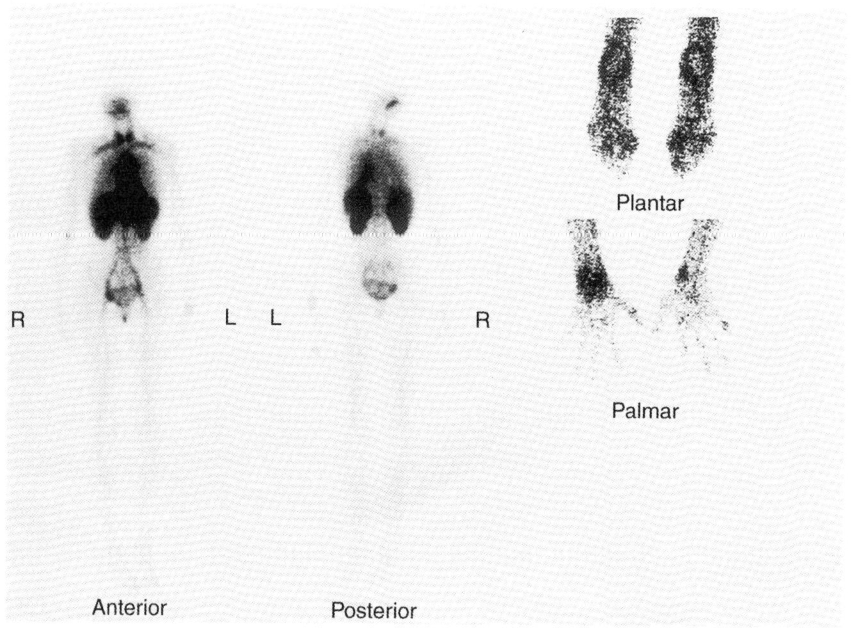

QUESTIONS

Q1. What does the study show?
Q2. What does this mean?

ANSWERS

Q1. Marked uptake in the wrists, some of the finger joints, the metatarso-phalangeal joints and the ankles.
Q2. Active rheumatoid arthritis.

TEACHING POINT

Tc HIG scintigraphy is a promising objective method to evaluate the presence of active inflammatory arthropathy, although its role requires further investigation.

Case 70

HISTORY

A 56-year-old woman presented with an unexplained painful left hip. A two-phase bone scan was performed.

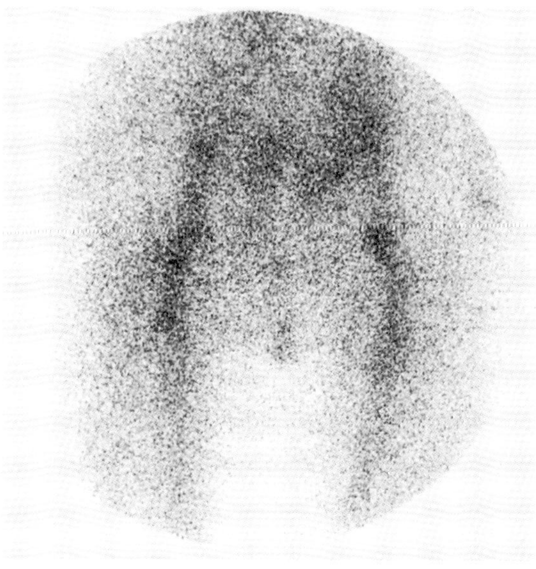

Blood pool.

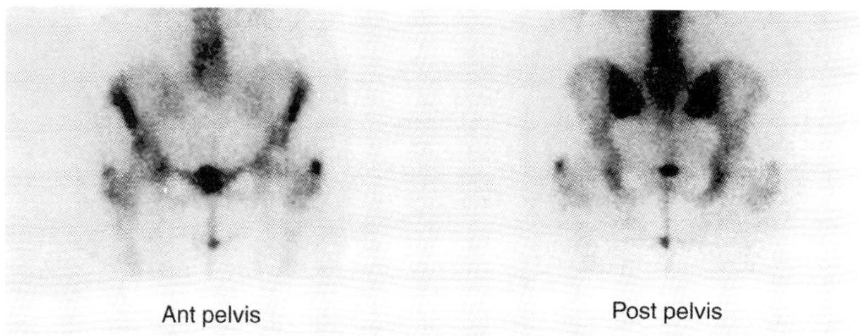

Static images.

QUESTIONS

Q1. What are the abnormal findings?
Q2. What is the diagnosis?

ANSWERS

Q1. Focal uptake in the left greater trochanter on blood pool and static images.

Q2. Greater trochanter bursitis.

Case 71

HISTORY

A 13-year-old boy presented with right hip pain. A two-phase bone scan was performed.

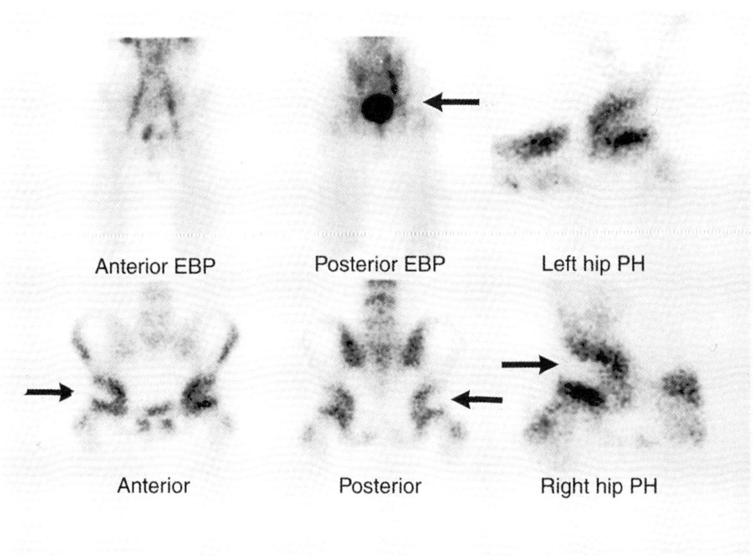

QUESTIONS

Q1. Describe the arrowed abnormality.
Q2. What is the diagnosis?

ANSWERS

Q1. Absent uptake on blood pool, static and pinhole (PH) images, right femoral head.

Q2. Appearances are typical of Perthes' disease.

Case 72

HISTORY

A 35-year-old woman with known Crohn's disease had continued abdominal pain and diarrhoea. She was already on oral prednisolone therapy. A labelled white cell scan was performed. Images are shown at one and three hours post-injection.

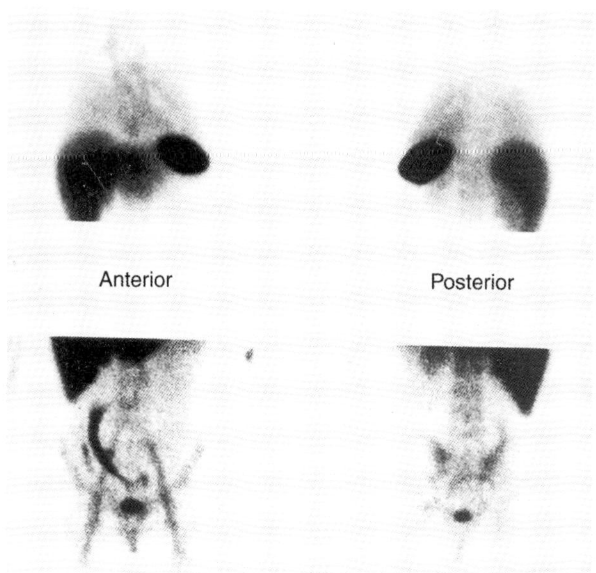

One hour.

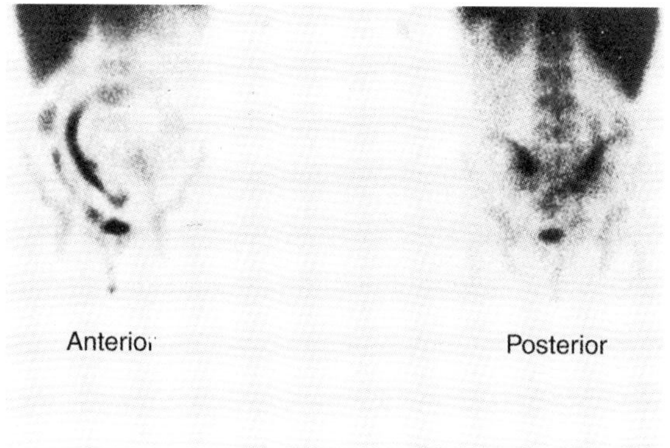

Three hours.

QUESTIONS

Q1. With what were the white cells labelled?
Q2. How does the result help the management of the patient?

ANSWERS

Q1. Technetium 99m hexamethyl propylene amine oxine (HMPAO). Bladder activity is not seen when indium is used as a label for white cell imaging.

Q2. There are multiple loops of inflamed small bowel. The study shows active disease and implies that the patient is not fully controlled.

TEACHING POINT

When using Tc HMPAO for white cell labelling, there is usually some renal and bladder activity, which should not be interpreted as abnormal.

FURTHER READING

Lantto, E., Jarvis, K., Kpetzela, I. *et al.* (1992) Technetium 99m hexamethyl propylene amine oxine leucocytes in the assessment of disease activity in inflammatory bowel disease. *Eur. J. Nuc. Med.*, **19**, 14–18.

Case 73

HISTORY

A 13-year-old girl presented with a painful right leg and pyrexia. On examination, her lower leg was swollen and tender. An urgent two-phase bone scan was requested. Blood pool and delayed static images of the lower legs were obtained.

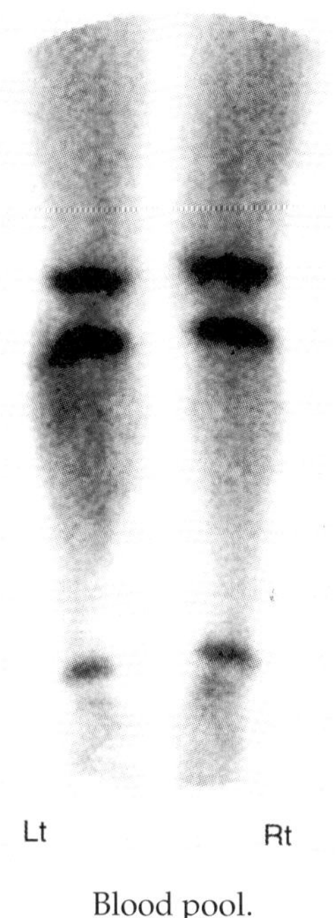

Blood pool.

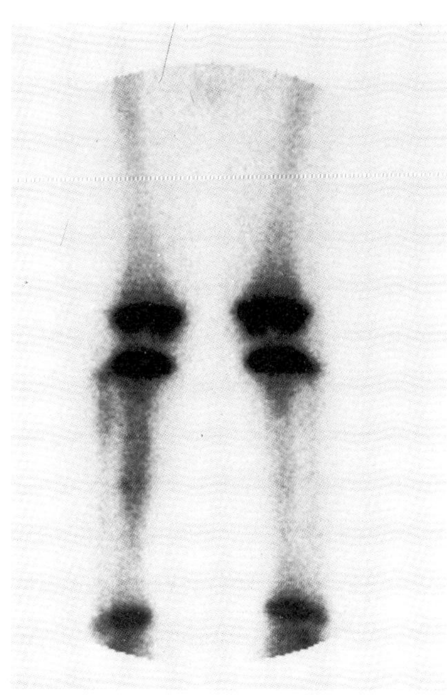

Delayed static image, anterior.

QUESTIONS

Q1. Describe the abnormalities.
Q2. What is the diagnosis?

ANSWERS

Q1. Increased activity on blood pool and delayed images right upper tibia.
Q2. Osteomyelitis.

TEACHING POINT

A patient with suspected acute osteomyelitis and normal x-rays will have the diagnosis detected with approximately 95% sensitivity and specificity with the two-phase bone scan.

FURTHER READING

P.J. Ryan and Fogelman, I. (1994) The role of nuclear medicine in orthopaedics. *Nuc. Med. Comm.*, **15**, 341–60.

Case 74

HISTORY

A 72-year-old man presented with possible myelomatosis. A bone marrow aspirate produced a dry tap. A nuclear medicine scan was performed to investigate the bone marrow.

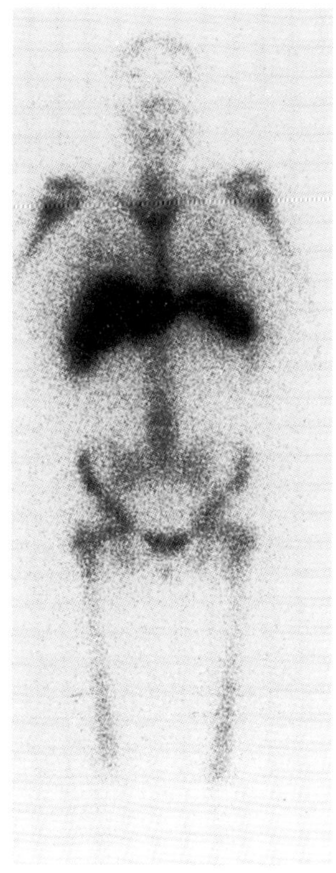

Whole body image.

QUESTIONS

Q1. What scan was performed?
Q2. What does it show?
Q3. How could this help in this clinical situation?

ANSWERS

Q1. Bone marrow scan, performed with ^{99m}Tc sulphur colloid.
Q2. Bone marrow hyperplasia.
Q3. The bone marrow scan can be used to show the distribution of marrow to enable the selection of the site for sampling where there is difficulty.

TEACHING POINTS

Bone marrow scans are also useful for:

- evaluation of patients who have had an equivocal bone scan for skeletal metastases
- detection of myeloma metastases
- detection of marrow infarction, as in sickle cell anaemia.

FURTHER READING

Reske, S.N. (1991) Recent advances in bone marrow scanning. *Eur. J. Nuc. Med.*, **18**, 203–21.

Case 75

HISTORY

A 25-year-old man was involved in a road traffic accident. He complained of left-sided loin pain. Ultrasound of the kidneys was normal. A DTPA study with frusemide was performed.

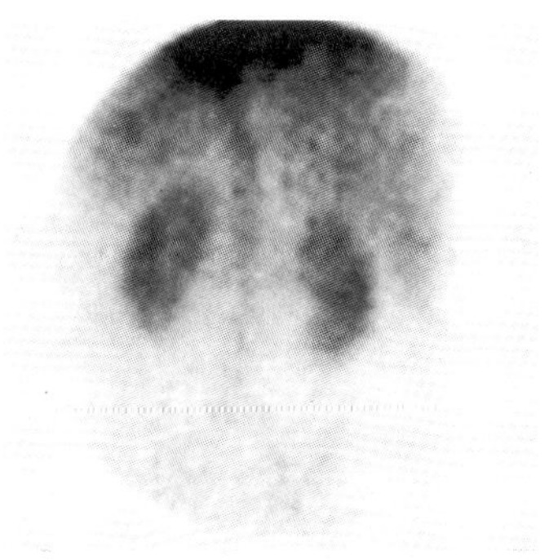

Zero to 30 seconds post-frusemide.

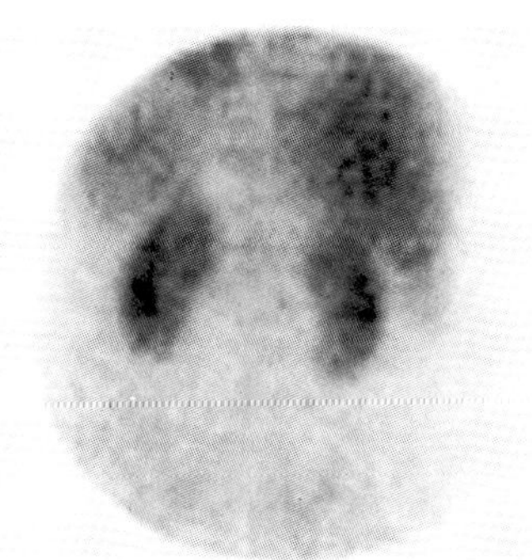

One minute post-frusemide.

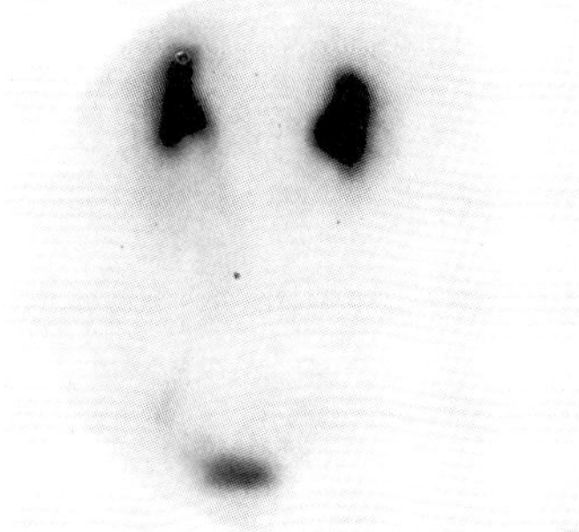

15 minutes post-frusemide.

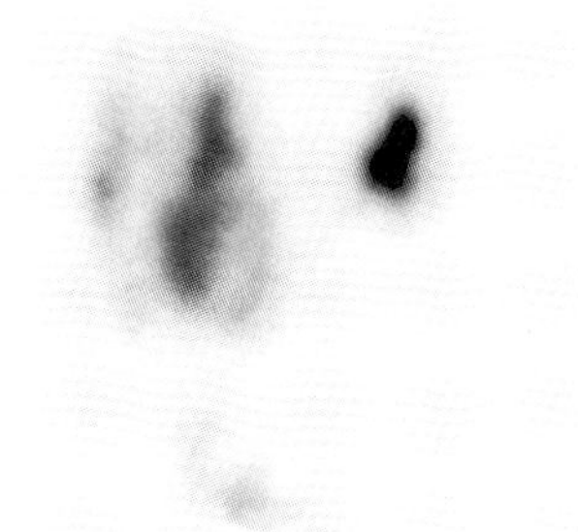

30 minutes post-frusemide.

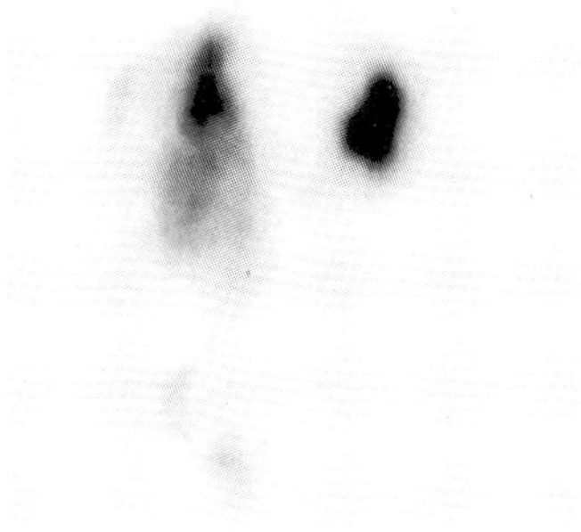

Two hours post-frusemide.

QUESTIONS

Q1. Describe the study.
Q2. What is the diagnosis?
Q3. Was the test the most appropriate?

ANSWERS

Q1. The left kidney is slightly enlarged but is well perfused and functions normally. At 15 minutes, a collection of tracer develops below the left kidney and becomes more apparent on the later images.

Q2. Urine leak.

Q3. A DTPA study is extremely sensitive for the detection of urine leakage and is the appropriate first-line investigation.

Case 76

HISTORY

A 17-year-old male was involved in a fight and fell on his left hand. X-rays were normal but there was a suspicion of a scaphoid fracture. A bone scan was performed. Blood pool, static and registered images of the hands and wrists are displayed.

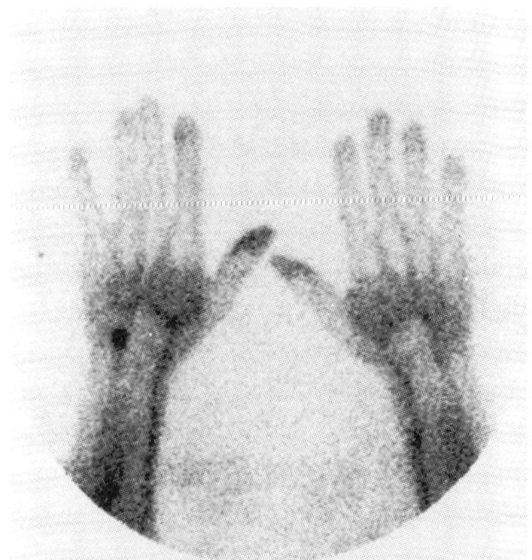

Blood pool.

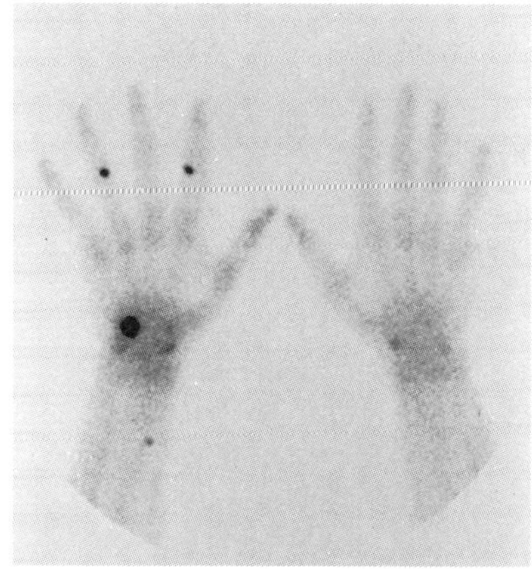

Static image.

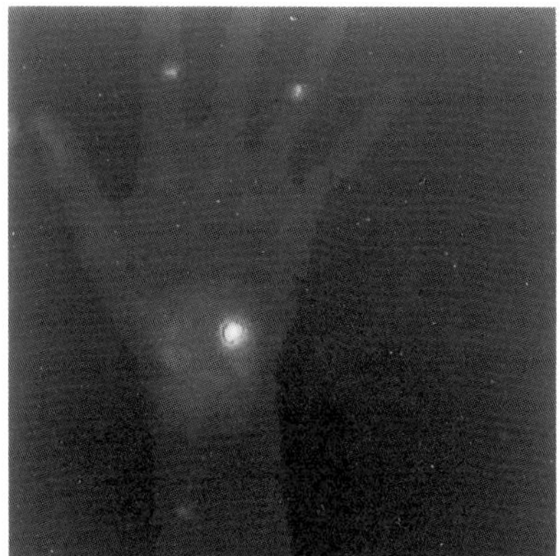

Registered image.

QUESTIONS

Q1. Describe the abnormality.
Q2. What is the diagnosis?

ANSWERS

Q1. There is increased blood pool and static activity of the ulnar aspect in the left wrist. The registered image localizes activity to the head of the hamate.

Q2. Fractured head of hamate.

TEACHING POINT

The bone scan is a very useful technique for diagnosing occult wrist fractures. Registration of the x-ray and scan allows more precise localization of the abnormal activity.

FURTHER READING

Mohammed, A.P., Ryan, P.J., Lewis, M. *et al.* (1993) Registration bone scan in the management of wrist pain. *Nuc. Med. Comm.*, **14**, 261.

Case 77

HISTORY

A 65-year-old man with known carcinoma of the prostate developed extensive bone pain and in particular low back pain. He had a rising prostate-specific antigen (PSA) and it was felt that he had now developed bony metastases and was escaping hormonal control. A whole body bone scan was performed.

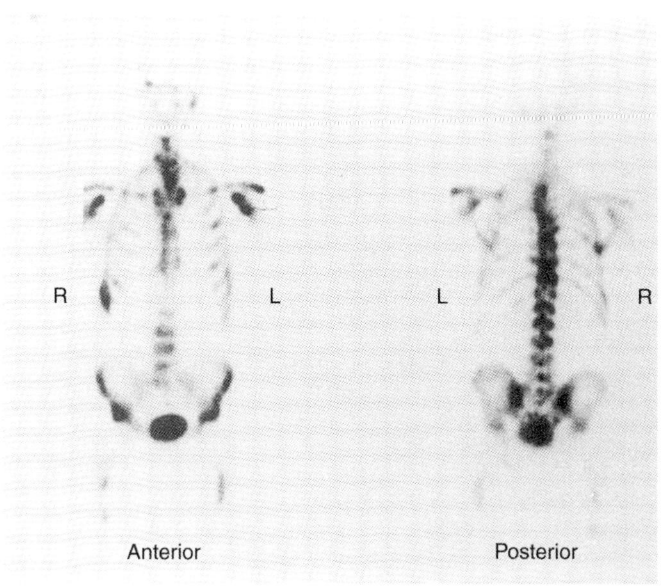

QUESTIONS

Q1. What is shown on the scan?
Q2. How may this guide management?

ANSWERS

Q1. Metastatic super scan. There are widespread skeletal metastases. The kidneys are not visualized and normal bone is barely visualized.

Q2. The patient would be usefully treated with local radiotherapy combined with radionuclide therapy (e.g. ^{89}strontium). The radionuclide therapy will prolong the period before further radiotherapy is required.

TEACHING POINT

Radiostrontium therapy is particularly useful for the palliative treatment of widespread pain caused by bony metastases and for concurrent treatment with radiotherapy where there is a single site of pain but multiple metastases on bone scan.

FURTHER READING

Porter, A.T., McEwan, A.J.B., Powe, J.E. *et al.* (1993) Results of a randomised phase III trial to evaluate the efficiency of Strontium 89 adjuvant to local field external beam irradiation in the management of endocrine resistant metastatic prostate cancer. *Int. J. Rad. Oncol. Biol. Phys.*, **25**, 805–13.

Case 78

HISTORY

A 25-year-old woman presented with thyrotoxicosis. There were no eye signs. A thyroid scan was performed with pinhole images of the thyroid.

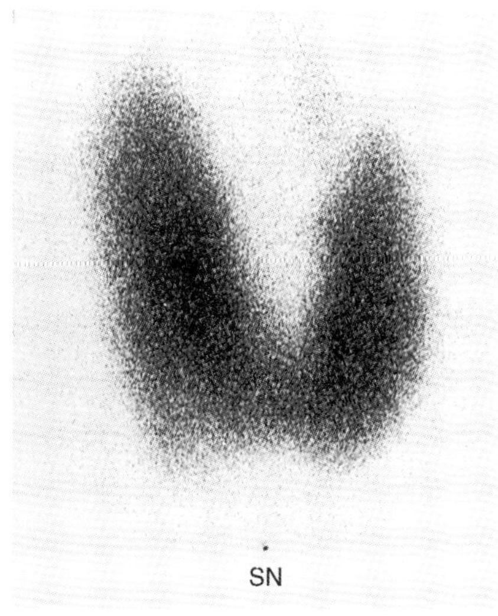

Thyroid scan.

QUESTIONS

Q1. What does the scan show?
Q2. How does this aid management of the patient?

ANSWERS

Q1. Diffuse tracer uptake with a homogeneous pattern in an enlarged thyroid. There is some asymmetry in lobe size, right more than left. There is some activity above the left lobe, which most probably represents a pyramidal lobe.

Q2. This pattern is typical for Graves' disease and is usually associated with elevated 20 minute uptake of tracer. Provided that the patient can be controlled with medical therapy, it is worth while treating for one to two years as the condition may resolve spontaneously.

Case 79

HISTORY

A 30-year-old woman with Crohn's colitis was continuing to have diarrhoea despite treatment with sulphasalazine, rectal steroids and oral prednisolone, 5 mg a day. An HMPAO white cell scan was performed to assess the extent of her disease. Images of the pelvis are displayed from one and three hours after re-injection.

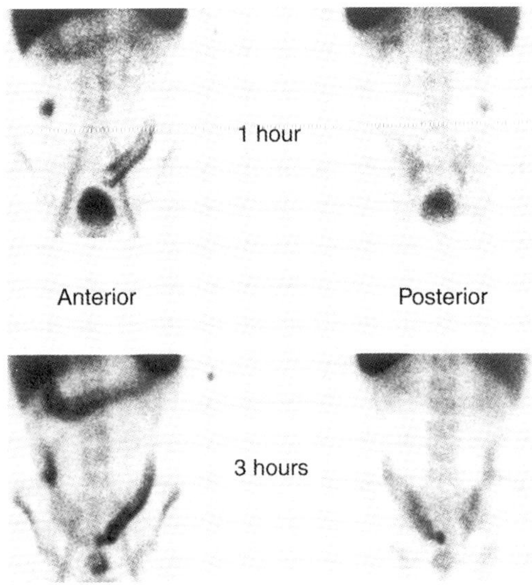

QUESTIONS

Q1. Describe the abnormalities.
Q2. How does this help the physician?

ANSWERS

Q1. There is activity on the one- and three-hour images involving the large bowel. This involves three separate areas on the one-hour image, but most of the large bowel appears involved on the delayed three-hour image.

Q2. The patient has extensive disease and will need an increase in systemic steroid therapy, or surgery. The small bowel has been shown to be unaffected on this study.

TEACHING POINT

The most useful role for white cell imaging in inflammatory bowel disease is to determine whether there is active inflammation and to define the extent of the disease. It is particularly useful for distinguishing symptoms due to inflammation from those due to mechanical obstruction. It is especially worth while in children because the test is better tolerated than barium studies and has a lower radiation dose.

FURTHER READING

Lantto, E., Jarri, K., Krekela, I. *et al*. (1992) Technetium 99m hexamethyl propylene amino oxine leucocytes in the assessment of disease activity in inflammatory bowel disease. *Eur. J. Nuc. Med.*, **19**, 14–18.

Case 80

HISTORY

A 70-year-old man developed persistent pain following the insertion of a cementless right knee prosthesis two years earlier. A bone scan followed by an HIG study was performed. Anterior blood pool, delayed bone images and three-hour post-injection HIG images are displayed.

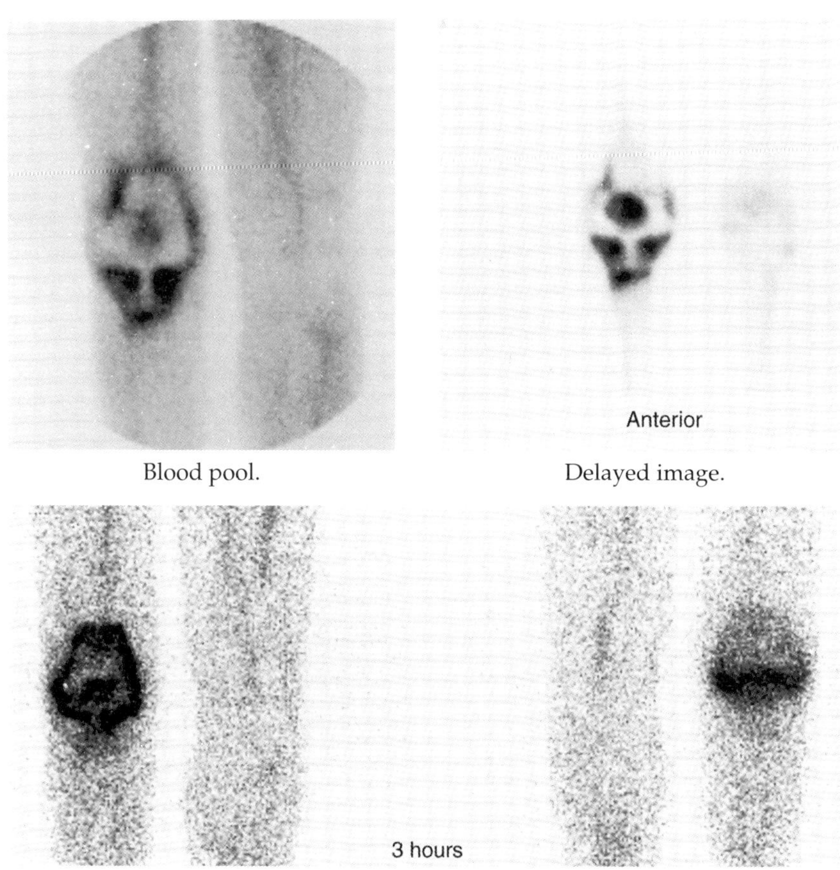

Blood pool.

Delayed image.

HIG study. Left, anterior image; right, posterior image.

QUESTIONS

Q1. Describe the abnormalities.
Q2. What is the diagnosis?

ANSWERS

Q1. There is increased blood pool and static activity on the bone scan that involves the patellar, tibial and femoral component of the left knee prosthesis. There is HIG activity in all the components of the prosthesis.

Q2. There is marked HIG activity involving the femoral component of the left knee prosthesis at the site of activity on the bone scan. This is strongly suggestive of the presence of infection.

TEACHING POINT

1. Knee prostheses are difficult to assess with a bone scan. A negative study essentially excludes infection, but a positive study may persist for years after surgery. A WBC scan may be necessary to clarify the situation.
2. Tc HIG is an alternative to white cell labelling for the detection of infection, although it is less sensitive. The mechanism of uptake is uncertain and its role is currently undefined. It may be particularly useful where sufficient blood cannot be obtained for white cell labelling, such as in children.

FURTHER READING

Serafini, A.N., Garty, J., Vargas-Cuba, R. *et al.* (1991) Clinical evaluation of a scintigraphic method for diagnosing inflammation/infection using indium 111 labelled non specific IgG. *J. Nuc. Med.*, **32**, 2227–32.

Case 81

HISTORY

A 64-year-old man of Indian ethnic background presented with back pain. He had a raised ESR of 90 mm/h; x-rays of the spine were normal. A whole body bone scan was performed.

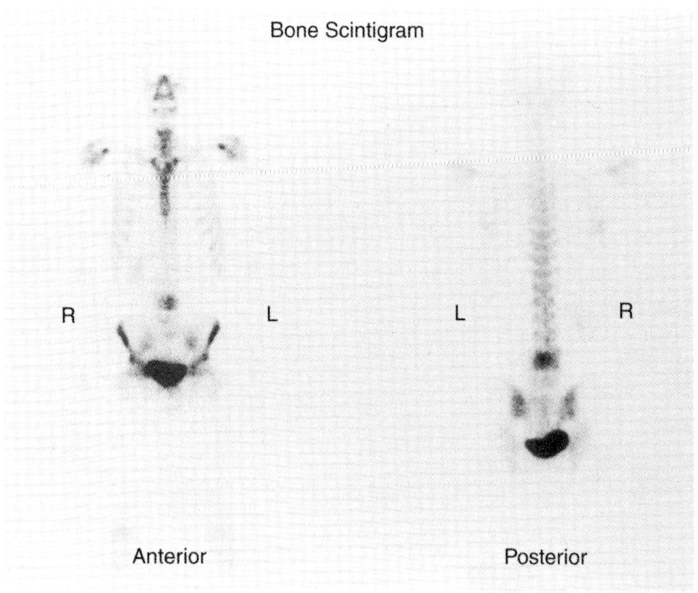

QUESTIONS

Q1. What is the abnormality?
Q2. What is the differential diagnosis?

ANSWERS

Q1. Increased activity at L4.
Q2. Single vertebral lesions can be due to many causes. For example:
- metastasis
- osteomyelitis
- fracture
- myeloma
- spondylolisthesis
- spondylosis

In this case the cause was spinal tuberculosis.

TEACHING POINT

Focal spinal lesions are non-specific. The scan may, however, provide clues. In this case, the activity is intense compared with other vertebrae and unlikely to be due to degenerative diseases. It is focal and the height of the vertebra appears normal; this is against a fracture. Osteomyelitis typically develops from a discitis, and it is usual to see uptake in vertebrae either side of an involved disc.

In this case, with no prior history, a single metastasis would be most likely. However, the ethnic background raised the possibility of tuberculosis, which was diagnosed following a biopsy.

Case 82

HISTORY

A 35-year-old woman presented with thyrotoxicosis. She had no eye signs, nor a palpable goitre. A thyroid scan was performed.

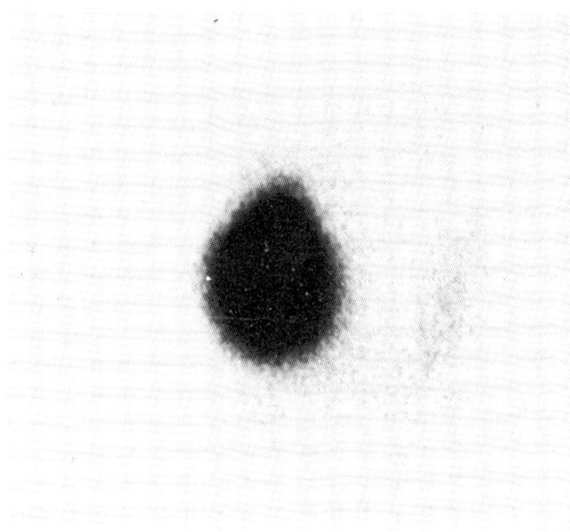

QUESTIONS

Q1. What does the scan show?
Q2. How does this influence the management of the patient?

ANSWERS

Q1. A solitary right-sided functioning thyroid nodule with barely visible left lobe. This suggests a toxic autonomously functioning thyroid adenoma.

Q2. The patient is best treated with early radio-iodine therapy, as a cure will not be achieved with medical therapy.

TEACHING POINT

The virtually absent left-sided thyroid tissue in this patient implies that the left lobe is non-functioning due to suppression of thyroid-stimulating hormone (TSH) by the thyroxine produced from the right-sided autonomously functioning adenoma. When the toxic adenoma is treated by radio-iodine, the left lobe will resume normal function. Note that on occasion a non-functioning left lobe could be due to a congenitally small left lobe or surgery to the left lobe. If in doubt, an ultrasound scan is worth while.

Case 83

HISTORY

A 56-year-old man developed an infected sinus above the right knee. He had previously had osteomyelitis of the lower femur. A combined bone and gallium scan was performed. The static bone images and 48-hour gallium images are displayed, both anterior views.

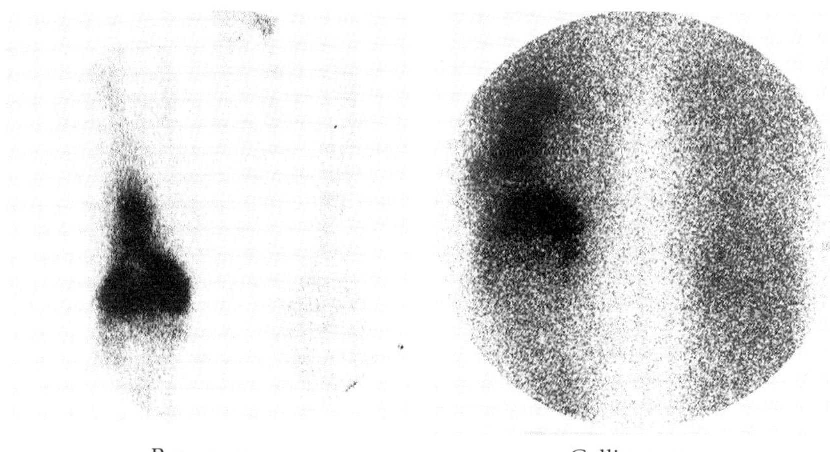

Bone scan. Gallium scan.

QUESTIONS

Q1. Describe the abnormalities.
Q2. What do they imply?

ANSWERS

Q1. On the bone scan there is increased uptake in the lower right femur. There is also gallium uptake. This is congruent with the bone scan at the level of the femoral condyle, but not congruent above this.

Q2. The increased activity on bone and gallium images at the head of the femoral condyle suggest active osteomyelitis. Gallium uptake above this may be in soft tissue rather than bone as it is not congruent with the bone scan.

TEACHING POINT

For confident diagnosis of bone infection with combined gallium and bone scan imaging, the gallium uptake should be greater than that seen on the bone scan or within bone but incongruent to bone scan activity.

REFERENCE

Schauwecker, D.S., Park, H.M., Mock, B.H. *et al.* (1984) Evaluation of complicating osteomyelitis with Tc-99m MDP, In-111 granulocytes and Ga-67 citrate. *J. Nuc. Med.*, **25**, 849–53.

Case 84

HISTORY

A 50-year-old man with known non-Hodgkin's lymphoma developed right hip pain. X-rays showed a lytic region in the anterior ilium. A gallium scan was then arranged. A 48-hour anterior whole body view is displayed.

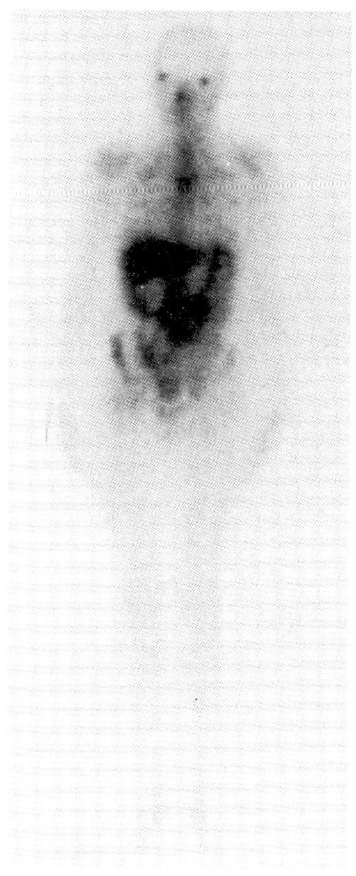

QUESTIONS

Q1. What is the abnormality?
Q2. What are the limitations of this study?

ANSWERS

Q1. There is increased tracer uptake in the right ilium supporting a lymphoma deposit as the explanation for the bone lesion on x-ray.

Q2. Gallium is a non-specific tracer and is taken up into a bone lesion because of a wide variety of causes.

Case 85

HISTORY

A 21-year-old runner developed right anterior knee pain. X-rays were unhelpful and MRI was normal. A bone SPECT study was performed. Tomographic images are displayed.

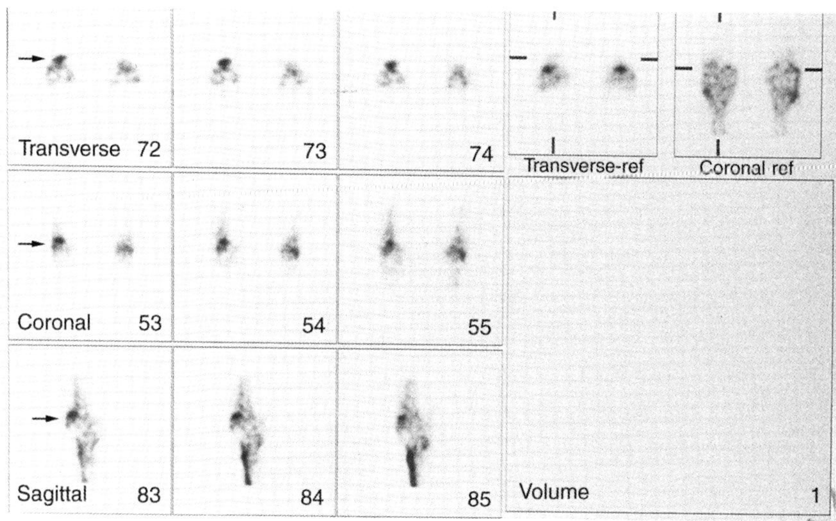

QUESTIONS

Q1. Where is the arrowed activity localized?
Q2. What is the diagnosis?

ANSWERS

Q1. The focal activity arrowed is localized to the superior margin of the patella.
Q2. Suprapatellar tendonitis.

TEACHING POINT

Bone SPECT is particularly valuable for examining the anterior compartment of the knee, although it can be used to examine the full range of skeletal knee pathology. Activity at the superior margin of the patella is the typical appearance of suprapatellar tendonitis.

Case 86

HISTORY

A 73-year-old man presented with thyrotoxicosis. There were no eye signs, but he had a palpable goitre on the right side only. A thyroid scan was performed.

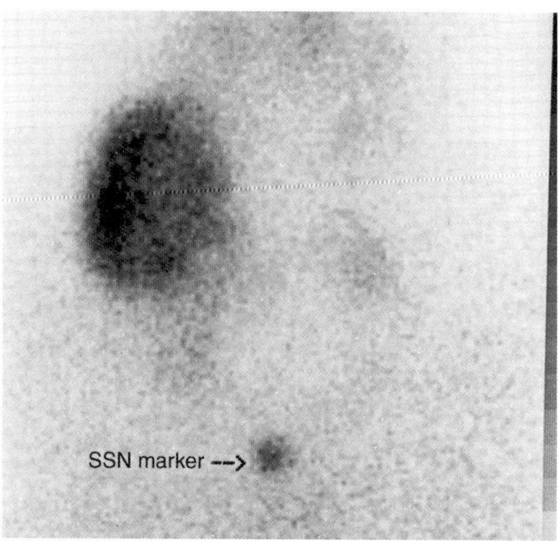

QUESTIONS

Q1. What abnormalities are visualized?
Q2. What treatment is appropriate?

ANSWERS

Q1. The thyroid lobes are markedly abnormal. On the right there is a nodule of increased activity, with an area of lesser uptake below. On the left, the gland is non-functioning apart from small areas of functioning tissue. The scan is of a multinodular goitre with a nodule on the right that is probably becoming autonomous. The scan appearances are those of Plummer's disease.

Q2. Early radio-iodine therapy because, like a solitary autonomously functioning adenoma, Plummer's disease will not settle with medical therapy. Higher doses of radio-iodine are commonly prescribed in Plummer's disease than in Graves' disease.

Case 87

HISTORY

A 60-year-old woman presented with watering of the right eye. The cause was clinically uncertain and the nasolacrimal duct was patent. A dacrocystogram was performed. Anterior images are displayed

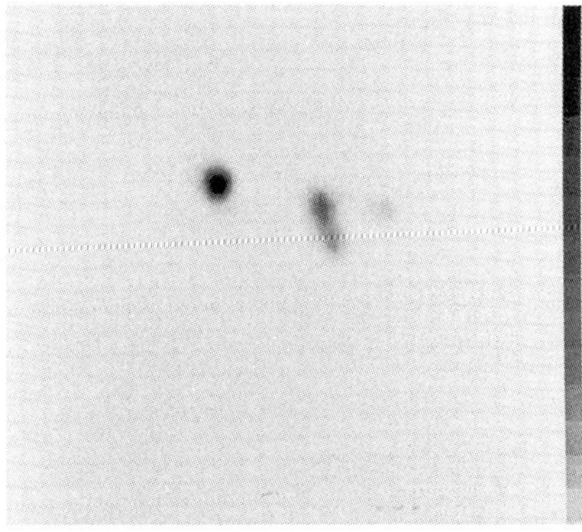

Five-minute image.

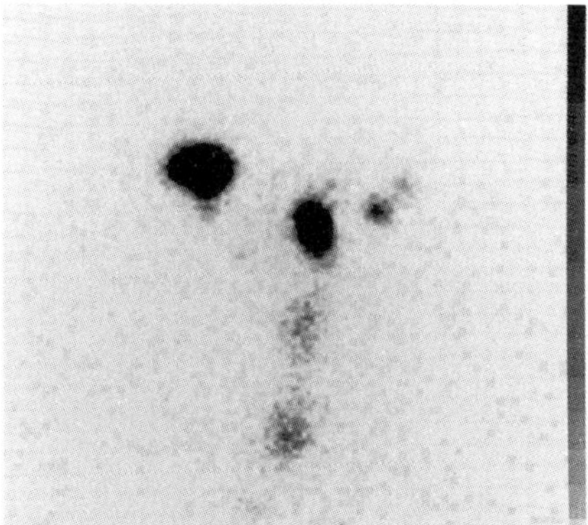

30-minute image.

QUESTIONS

Q1. Describe the abnormality on the right side.
Q2. How does this help the ophthalmic surgeon?

ANSWERS

Q1. Tracer is placed under both lower eyelids. On the left it enters the punctum and passes into the nasolacrimal sac and duct, and eventually enters the nose. On the right, tracer does not pass trough the punctum but remains pooled in the eye.

Q2. The hold-up on the right is identified as a failure to enter the punctum and relates to failure of the right eye to deliver lacrimal fluid to the punctum.

Case 88

HISTORY

A 70-year-old woman presented with a urinary tract infection. An ultrasound scan showed a left hydronephrosis. A MAG 3 study was performed. Images obtained at 60 secs, 2 mins, 5 mins, 10 mins, 15 mins, 20 mins and 25 mins following injection and pre- and post-micturition images are shown. The response of the kidneys is shown on the time–activity curve (left kidney bold and right kidney faint).

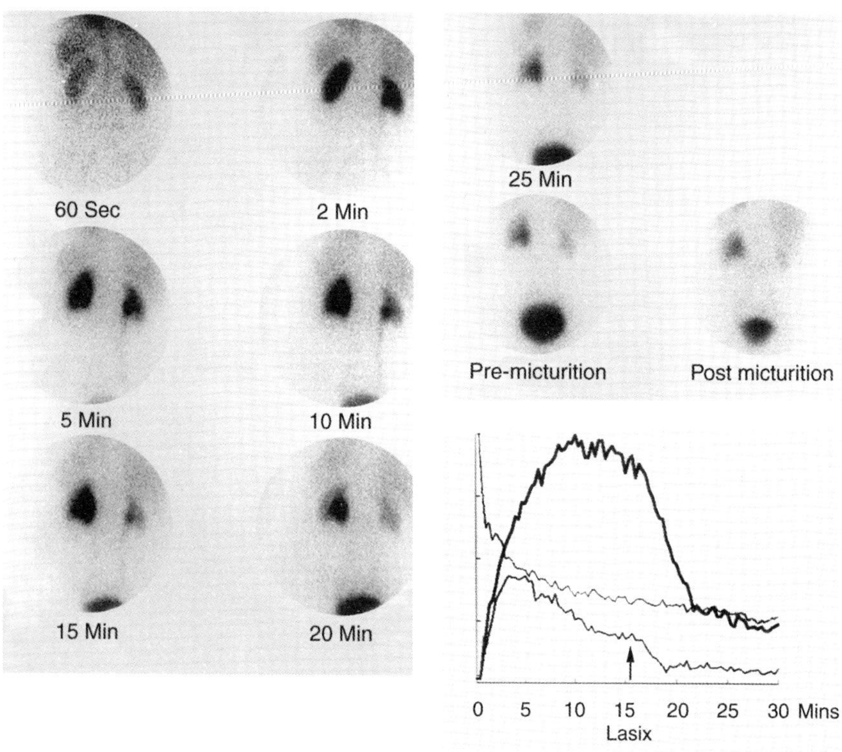

QUESTIONS

Q1. Describe the abnormalities in this study.
Q2. What does this imply?

ANSWERS

Q1. An enlarged left kidney. This has good function but accumulates tracer into a dilated renal pelvis. There is good drainage however following frusemide.

Q2. Dilated non-obstructed left kidney.

TEACHING POINT

A DTPA or MAG 3 study is extremely helpful in distinguishing PUJ obstruction from a dilated non-obstructed renal pelvis. Brisk drainage following frusemide excludes significant obstruction. In patients with equivocal studies quantitation of the frusemide washout curve can be helpful. Prolonged parenchymal transit is also a feature of obstructive nephropathy and its measurement by deconvolution analysis can help with difficult cases.

Case 89

HISTORY

A 56-year-old man presented with a left-sided neck swelling. On examination, he had a left-sided goitre. A thyroid scan was performed followed by a thallium scan (three hours post-injection).

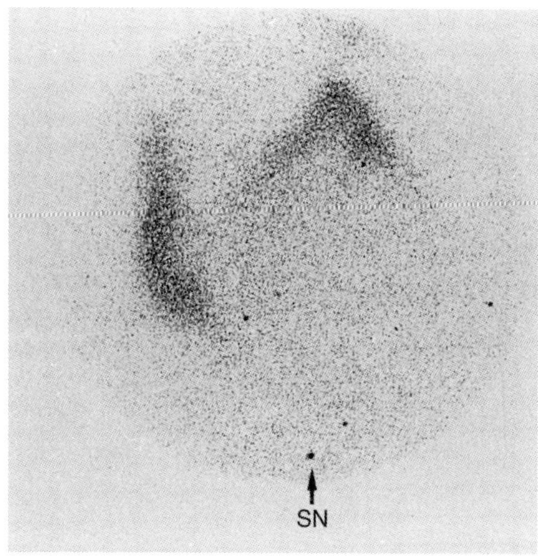

Thyroid scan.

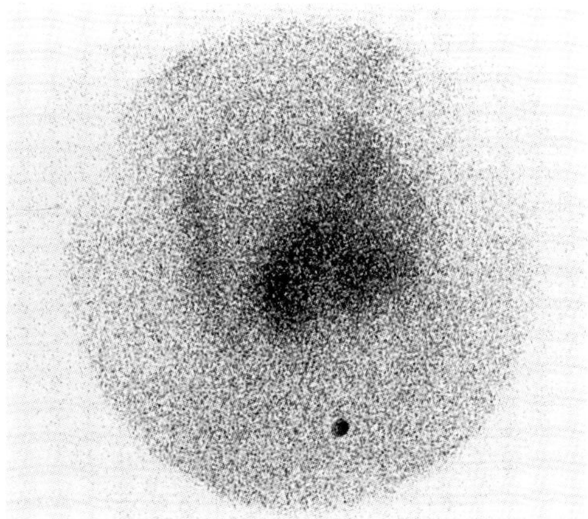

Thallium scan of the neck.

QUESTIONS

Q1. What is the thyroid scan abnormality?
Q2. What does the thallium scan show?

ANSWERS

Q1. A non-functioning (cold) left thyroid nodule.
Q2. Thallium uptake at the site of the cold nodule. In this patient, the diagnosis was medullary carcinoma of the thyroid.

TEACHING POINT

Non-functioning or hypofunctioning thyroid nodules may be malignant and require further investigation. This will usually be with fine needle aspiration cytology (FNA). Where FNA is equivocal or technically difficult, a thallium scan is valuable. Increased uptake on delayed thallium images has a high sensitivity for thyroid carcinoma. Uptake less than the remainder of the thyroid on delayed images makes carcinoma unlikely.

Case 90

HISTORY

A 70-year-old man developed a painful left hip. A cemented prosthesis had been inserted 10 years previously. A bone scan was performed. The dynamic images were normal and the anterior image is displayed.

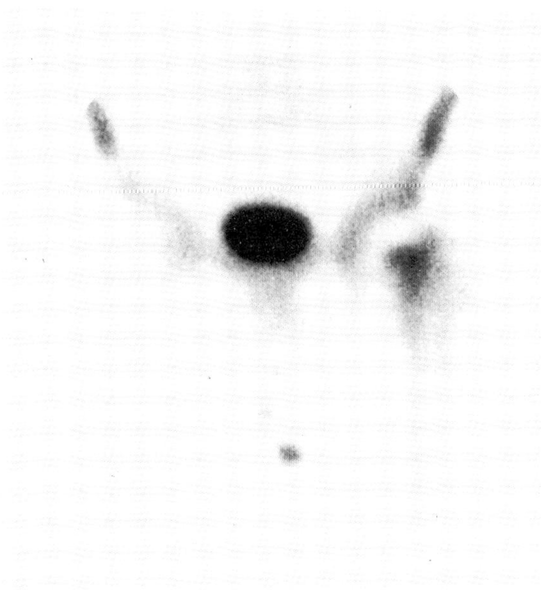

QUESTIONS

Q1.What abnormality is demonstrated?
Q2. What is the likely diagnosis?

ANSWERS

Q1. Increased activity in the left lesser trochanter.
Q2. Loosening femoral component of the hip prosthesis.

TEACHING POINT

The most common pattern of abnormality associated with loosening of the femoral component is tip activity. However, abnormal uptake is limited to the lesser trochanter in some cases, and this is the second most common pattern of abnormality.

FURTHER READING

Rubello, D., Borsato, N., Chierichetti, F. *et al*. (1995) Three phase bone scintigraphy pattern of loosening in uncemented hip prosthesis. *Eur. J. Nuc. Med.*, **22**, 299–301.

Case 91

HISTORY

A 20-year-old woman was investigated for recurrent urinary tract infection. A DMSA study was performed. A posterior three-hour image is displayed.

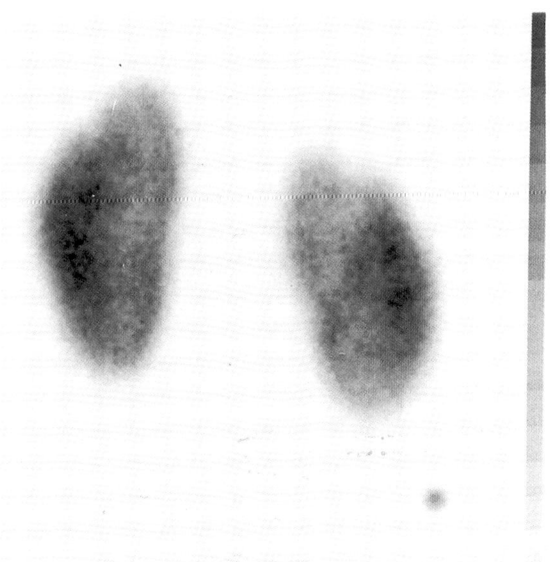

QUESTIONS

Q1. Describe the abnormalities
Q2. How should these be interpreted?

ANSWERS

Q1. There is cortical irregularity in both upper poles.
Q2. Scan findings are typical for renal scarring.

TEACHING POINT

Defects on DMSA scanning can be due to pyelonephritis, although the typical pattern in this diagnosis is focal loss of activity without cortical irregularity.

Case 92

HISTORY

A 30-year-old man was investigated for recurrent urinary tract infection. He was known to have reflux on micturating cystogram. A DMSA study was performed. The posterior view is displayed.

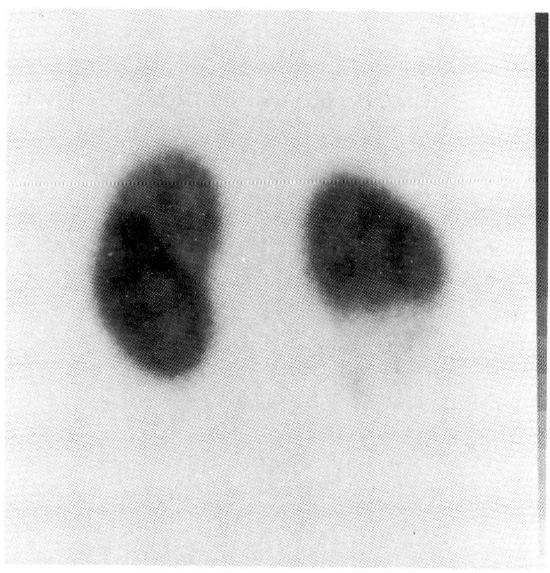

QUESTIONS

Q1. Describe the scan abnormality.
Q2. What is the differential diagnosis?

ANSWERS

Q1. A large photon-deficient lesion in the left lower pole.
Q2. The extensive defect in the lower pole while the remainder of the kidney is normal is suggestive of pathology within a duplex kidney. Scan changes could be due to an obstructed lower pole. Alternatively, there could be scaring and dilated collecting systems due to reflux in the lower pole.

TEACHING POINT

Lower pole duplex kidneys are more prone to obstruction than reflux. Large photon deficient areas can also be due to cysts and tumour and further investigation to detect or exclude these is required.

Case 93

HISTORY

A 50-year-old presented with facial flushing and weight loss. Investigation revealed an elevated 5-hydroxyindoleacetic acid (5HIAA) level. An octreoscan was performed. A whole body image at 24 hours is displayed.

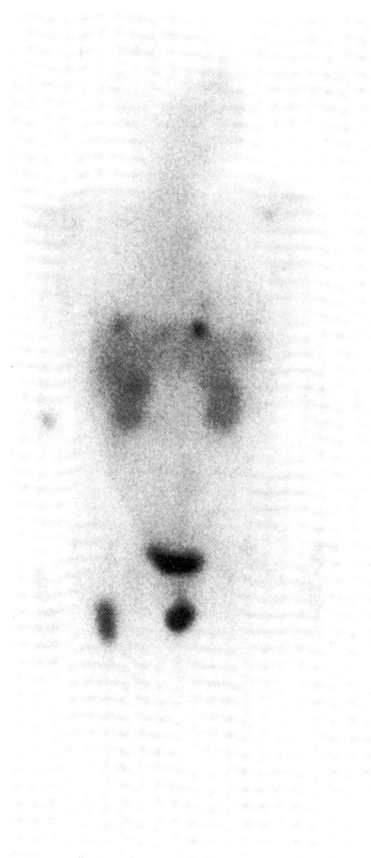

QUESTIONS

Q1. Describe the abnormalities.
Q2. What is the likely diagnosis?

ANSWERS

Q1. Multiple focal abnormalities are identified in the liver, chest, left shoulder and right femur.
Q2. Carcinoid syndrome with multiple metastases.

TEACHING POINTS

1. Octreotide is a somatostatin analogue, and when labelled with ^{111}In it can be used to detect tumours expressing somatostatin receptors.
2. Octreotide scintigraphy is the most sensitive test for carcinoid syndrome, for the detection of both the primary tumour and metastases. Liver metastases will be seen where CT is negative and whole body imaging gives the opportunity to detect metastases anywhere in the body.

Case 94

HISTORY

A 38-year-old man was investigated for poorly controlled hypertension. A baseline DTPA study was performed. Posterior images are displayed.

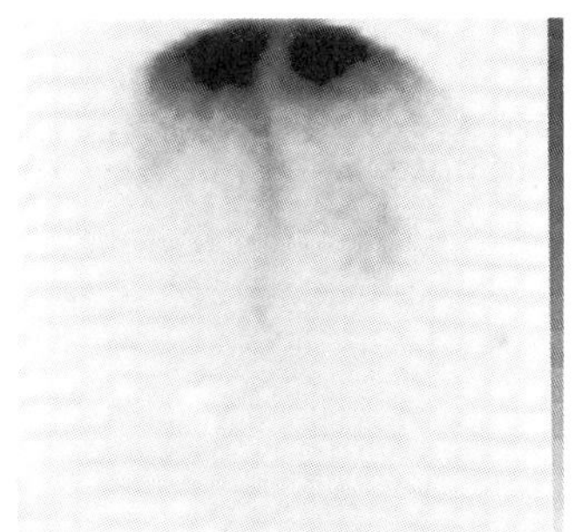

Zero to 30 seconds post-injection.

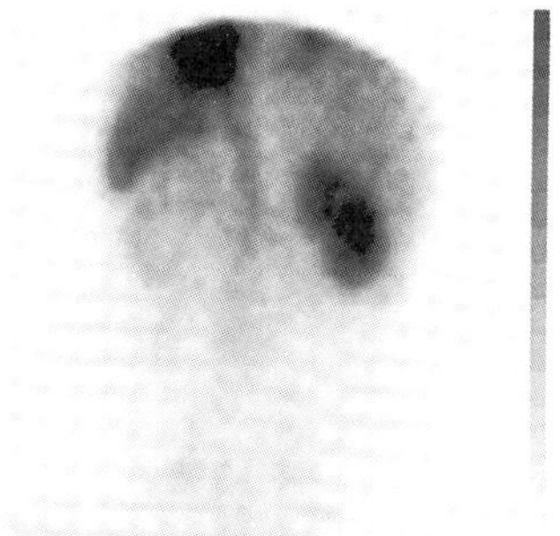

One minute post-injection.

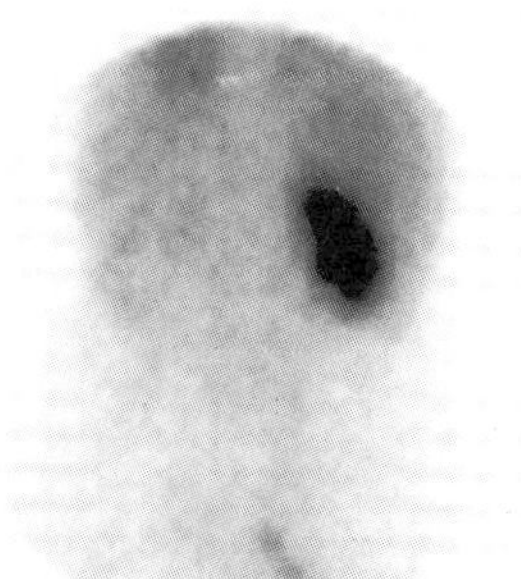

Five minutes post-injection.

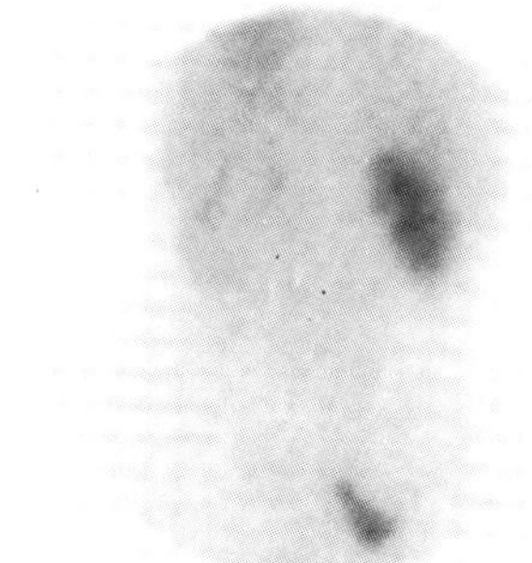

10 minutes post-injection.

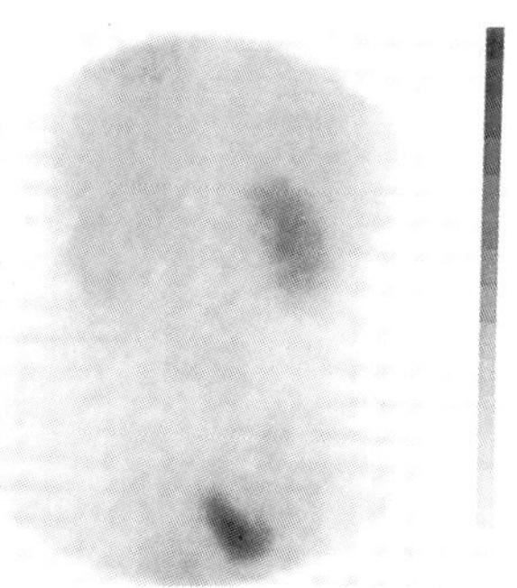

20 minutes post-injection.

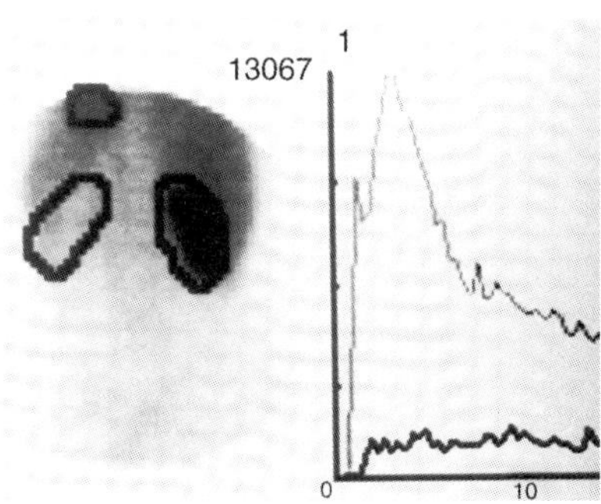

Time activity curve.

QUESTIONS

Q1. What does the scan show?
Q2. How should this be interpreted?

ANSWERS

Q1. Small poorly perfused, very poorly functioning left kidney.

Q2. There are many causes of the above findings, including severe renal artery stenosis. Further evaluation with ultrasound and/or an angiogram is appropriate.

TEACHING POINT

In a patient with suspected renovascular hypertension due to renal artery stenosis, a small poorly functioning kidney constitutes a high probability study. There is little value in performing a captopril study in this setting because there is minimal scope for demonstrating further decline in function.

Case 95

HISTORY

A 73-year-old woman presents with confusion and loss of short-term memory. Dementia is suspected. A nuclear medicine brain scan is performed. Transaxial views are displayed.

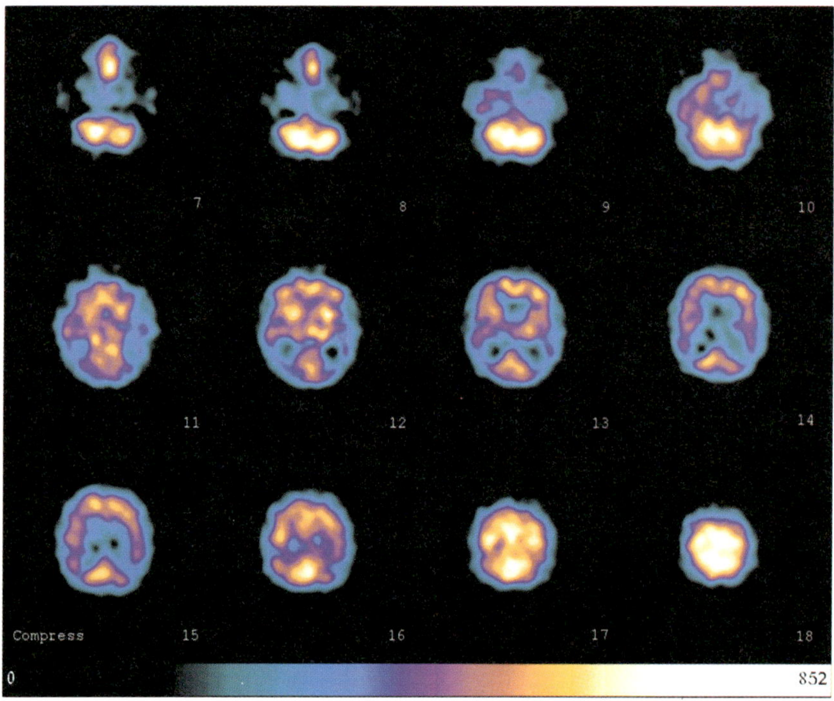

QUESTIONS

Q1. What is the scan?
Q2. What are the abnormalities?
Q3. What is the diagnosis?

ANSWERS

Q1. Tc HMPAO SPECT brain scan.
Q2. Reduced grey matter perfusion with generalized reduced tracer uptake except for the cerebellum, subcortical structures and occipital cortex. There is particularly reduced temperoparietal uptake.
Q3. Alzheimer's disease.

FURTHER READING

Bonte, F.J., Hom, J., Tintner, R. and Weiner, M.F. (1990) Single photon tomography in Alzheimer's disease and the dementia. *Sem. Nuc. Med.*, **XX**(199), 342–52.

Case 96

HISTORY

A 54-year-old man presents with atypical chest pain. A stress ECG was equivocal, showing 1 mm ST depression V6. A nuclear medicine study was performed.

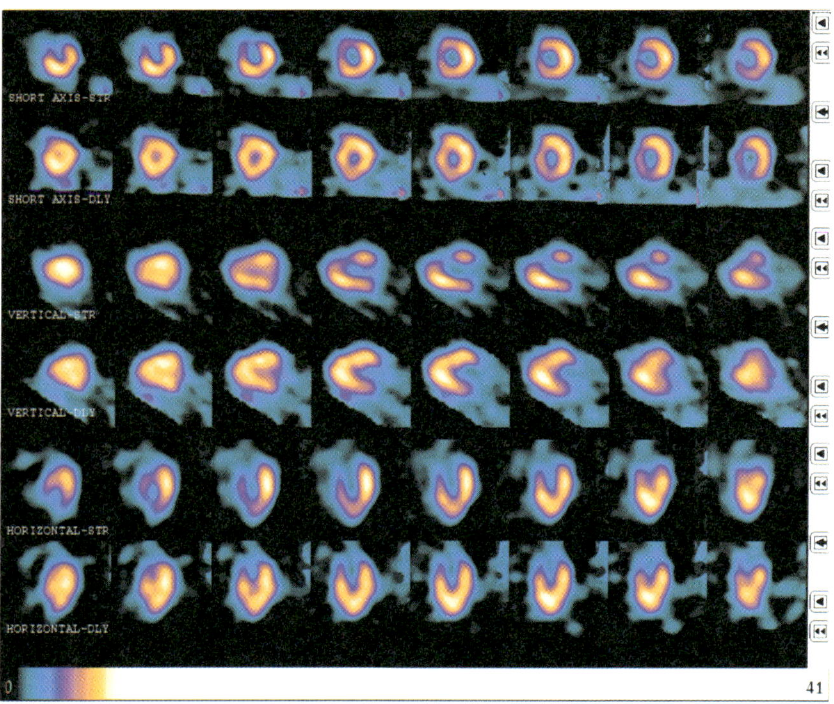

QUESTIONS

Q1. What is the scan?
Q2. What does it demonstrate?

ANSWERS

Q1. Exercise stress thallium scan.

Q2. There is impaired perfusion in the anterior wall on the stress study, but redistribution of tracer on delayed images. This demonstrates the presence of myocardial ischaemia in the left anterior descending artery territory.

TEACHING POINT

At peak stress, blood flow is diverted from stenosed vessels to healthy vessels, provoking ischaemia. Under resting conditions, tracer redistributes to ischaemic myocardium from normally perfused areas. Following stress, tracer washes out of healthy myocardium. This is markedly slower in the presence of ischaemia, and there is probably also 'wash in' of tracer to ischaemic sites from normally perfused areas.

FURTHER READING

Verani, M.S. (1992) 201 Tl myocardial perfusion imaging. *Curr. Op. Radiol.*, **3**, 797–809.

Case 97

HISTORY

A 56-year-old man was admitted with a small subendocardial myocardial infarct. He remained well but proceeded to have a thallium scan six weeks later.

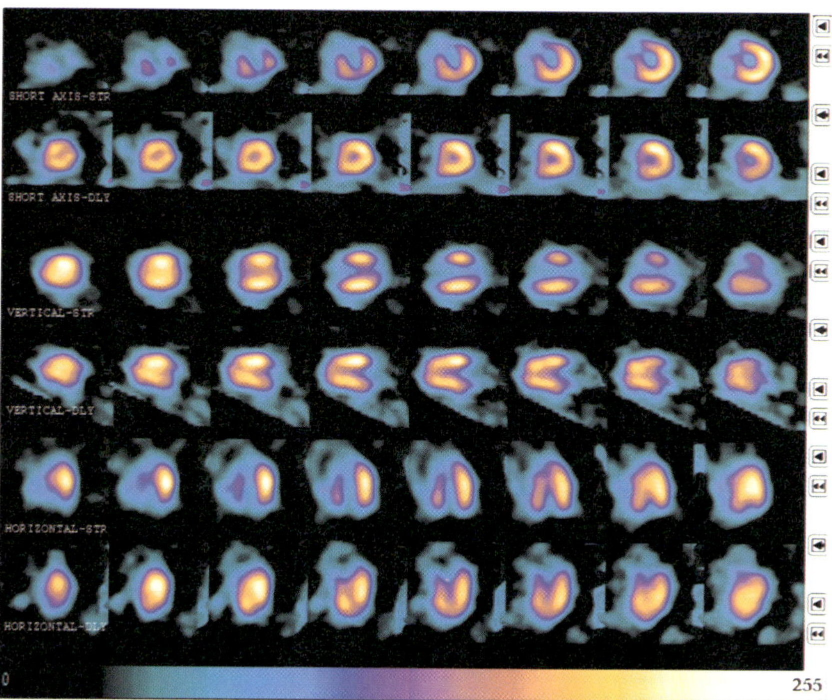

QUESTIONS

Q1. What does the scan show?
Q2. What does this mean?
Q3. Why was the scan performed?

ANSWERS

Q1. Impaired perfusion on stress images of the anterior wall, apex and septum. There is reperfusion on delayed images.

Q2. Myocardial ischaemia throughout much of the left anterior descending artery territory.

Q3. To determine the requirement for angiography post-myocardial infarct. Patients without evidence of myocardial ischaemia have a <1% chance per annum of further infarction. Those with evidence of ischaemia, as in this patient, have a 5% chance or more per annum of a further infarct.

FURTHER READING

Verani, M.S. (1994) Exercise and pharmacological stress testing for progression after acute myocardial infarction. *J. Nuc. Med.*, **55**, 716–20.

Case 98

HISTORY

A 70-year-old man presented with atrial fibrillation. No cause was found on routine check-up, and myocardial ischaemia was considered. An exercise stress electrocardiogram (ECG) could not be performed due to intermittent claudication. A dipyridamole thallium scan was performed.

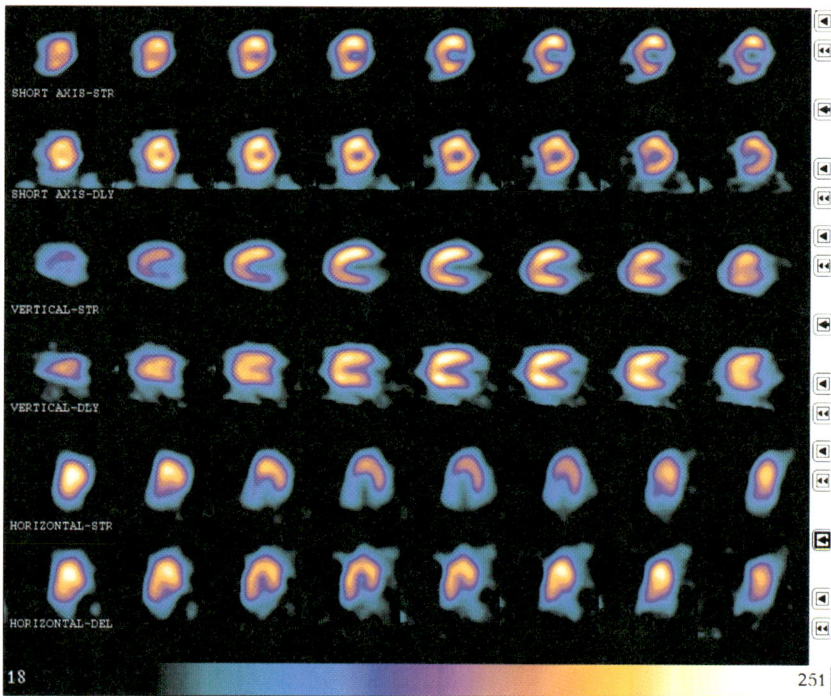

QUESTIONS

Q1. What does the scan show?
Q2. How does dipyridamole stress work?

ANSWERS

Q1. Reversible myocardial ischaemia in the left circumflex artery territory (lateral wall).

Q2. Dipyridamole is converted to adenosine, a potent coronary vasodilator. The effects on the coronary vessels are similar to those of exercise.

TEACHING POINT

Pharmacological stress as an alternative to exercise can be performed with dipyridamole, adenosine or dobutamine. Dipyridamole is the one used most widely. However, dobutamine should be used in asthmatic patients, in whom dipyridamole and adenosine are contraindicated.

REFERENCE

Pennell, D.J. (1994) Pharmacological cardiac stress: when and how? *Nuc. Med. Comm.*, **15**, 578–85.

Case 99

HISTORY

An 85-year-old man with a history of a previous cardiovascular accident presented with confusion and disorientation. Clinically he appeared to be developing dementia. A Tc HMPAO SPECT scan was performed.

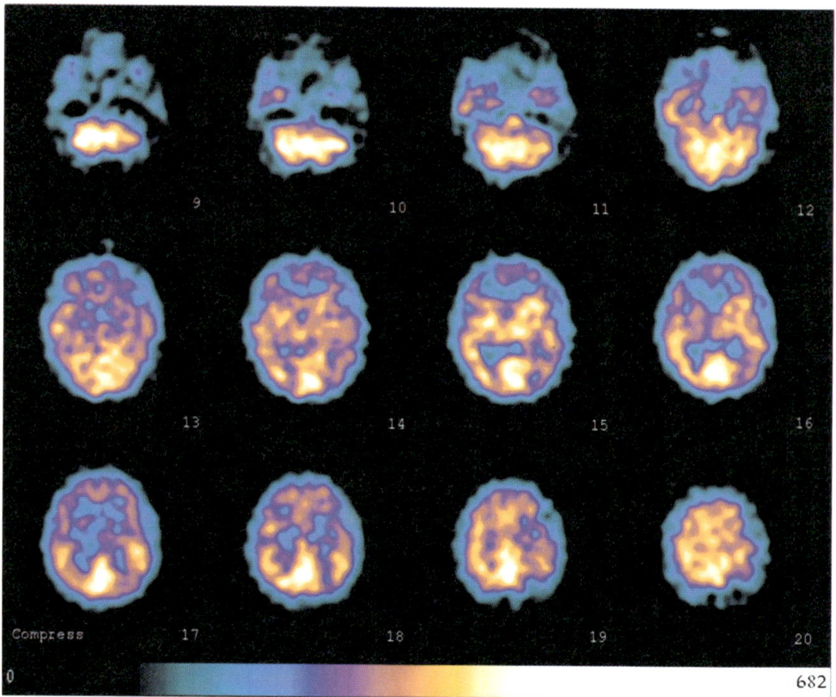

QUESTIONS

Q1. What does the scan show?
Q2. What does this imply?

ANSWERS

Q1. Multiple perfusion defects, with generally poor white matter perfusion.
Q2. Multi-infarct dementia.

Case 100

HISTORY

A 60-year-old man was considered for aortic aneurysm surgery. He had intermittent claudication and a past history of possible angina. A stress thallium scan was performed.

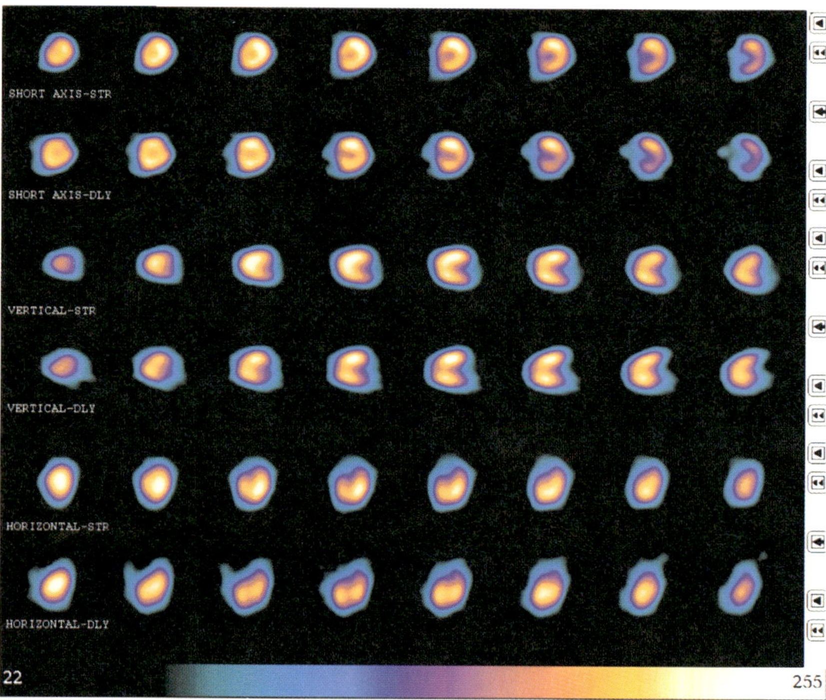

QUESTIONS

Q1. What does the scan show?
Q2. What does this imply?

ANSWERS

Q1. Normal perfusion of the myocardium with a small muscular septum.

Q2. A normal study implies a very low (<1%) chance of peri-operative ischaemia.

Index